A PRACTICAL GUIDE TO ENGINEERING, PROCUREMENT AND CONSTRUCTION CONTRACTS

PRACTICAL CONSTRUCTION GUIDES

Construction Claims
Current Practice and Case Management
Jeremy Hackett

Partnering and Collaborative Working
Edited by Rona Westgate, David Jones and David Savage

Construction Law and Management
Edited by Keith Pickavance

A Practical Guide to Successful Construction Projects
Arent van Wassenaer

A Practical Guide to Engineering, Procurement and Construction Contracts
Eric Eggink

For more information about this series, please visit:
www.routledge.com/Practical-Construction-Guides/book-series/PCG

A PRACTICAL GUIDE TO ENGINEERING, PROCUREMENT AND CONSTRUCTION CONTRACTS

ERIC EGGINK

informa law
from Routledge

First published 2020
by Informa Law from Routledge
2 Park Square, Milton Park, Abingdon, Oxon OX14 4RN

and by Informa Law from Routledge
52 Vanderbilt Avenue, New York, NY 10017

Informa Law from Routledge is an imprint of the Taylor & Francis Group, an informa business

British Library Cataloguing-in-Publication Data
A catalogue record for this book is available from the British Library

Library of Congress Cataloging-in-Publication Data
Names: Eggink, Eric, author.
Title: A practical guide to engineering, procurement and construction contracts / Eric Eggink.
Description: Abingdon, Oxon ; New York, NY : Routledge, 2020. | Series: Practical construction guides | Includes index.
Identifiers: LCCN 2019052757 (print) | LCCN 2019052758 (ebook) | ISBN 9780367344962 (paperback) | ISBN 9781003026372 (ebook)
Subjects: LCSH: Engineering contracts.
Classification: LCC K891.B8 .E34 2020 (print) | LCC K891.B8 (ebook) | DDC 624.068/7—dc23
LC record available at https://lccn.loc.gov/2019052757
LC ebook record available at https://lccn.loc.gov/2019052758

ISBN: 978-0-367-34496-2 (pbk)
ISBN: 978-1-003-02637-2 (ebk)

Typeset in Times New Roman
by Apex CoVantage, LLC

OUTLINE CONTENTS

CONTENTS

ACKNOWLEDGEMENTS

In 1995, when I was working as in-house legal counsel for Ballast Nedam International, our managing director Martin Weck approached me stating that the European International Contractors (EIC) organisation, of which he was a board member, had formed a working group that was going to undertake reviews and prepare comments on drafts of contract conditions to be published (in 1999) by the Fédération Internationale Des Ingénieurs-Conseils, also known as FIDIC. He asked me to take part in this working group reviewing the new, yet-to-be published, FIDIC Contract Conditions. I shall be forever indebted to him for asking me to take part in this endeavour. The working group consisted of a varying number (between seven and ten) of people, and sharing knowledge and experiences with colleagues from all over Europe has taught me more than I could ever have gotten from books. It was like having all these experiences myself, listening to their "war stories" and telling mine, while we were relating the text of the draft FIDIC Conditions to our experiences in the field. To all readers in a management role I say: if you are ever in a position to delegate one of the people you are responsible for to a working group of people discussing or conceiving contract conditions, do not hesitate to nominate them; it may take some time away but in one year's time they will gain the knowledge that would otherwise take them ten to 15 years to gain.

In 2003 I was again approached, this time by Robert Knutson, who was chairing an ICC working group to conceive contract conditions for an EPC/Turnkey Contract for Major Projects for the International Chamber of Commerce. Again, the experience of working with colleagues from all over Europe, relating contract condition texts to real-life situations and sharing war stories, was amazing and multiplied my knowledge of contractual issues in the construction industry by many times. When I left the civil engineering industry in 2007 and went into the oil and gas industry, where mostly bespoke contracts are used, I could make the transfer without batting an eye. Without these experiences such a transfer would have been much more difficult. Becoming familiar with standard conditions of contract takes a little time, maybe, but there is no such thing as familiarity with bespoke contracts, though one will discover that any particular Owner in the oil and gas industry often uses the same contract conditions for similar projects but then often changes certain provisions as they did not work so well in a recent project for them. With regard to the oil and gas industry: I owe a debt of gratitude to my colleague Michael Jackson of Technip, who taught me the intricacies of EPC/Turnkey contracting in the oil and gas industry.

Subsequently I had the honour of chairing, sometimes co-chairing, working groups within the ICC preparing model contracts for Joint Ventures and Consortia. I am indebted to all my colleagues in these working groups and owe them a debt of gratitude for their openness and courage to share their stories without fear of being ridiculed for stories where things went wrong. Yes, they were not all success stories as stories of failures are often more illustrative than stories of success.

Also, I must thank Arent van Wassenaer, who at the launch of his book *A Practical Guide to Successful Construction Projects* introduced me to his editor Amy Jones and asked me when I was going to finally write my book, and thereby planted the seed in my brain that has now grown into this book that you have in your hands; I also owe thanks to Amy Jones' and her staff's efforts.

Last but not least I owe a great debt of gratitude to my darling wife Celeste, who had to endure me spending hours, weeks and months in the attic bent over my laptop and writing, not giving her any time or attention, and who still would massage my cramped shoulders; by encouraging me, she helped to nurse the seed in my brain to come to full bloom.

May the fruit of this labour, the penning-down all lessons learnt in all the working groups I took part in, benefit your projects and may they all become a great success.

Eric Eggink
Rijnsburg, October 2019

INTRODUCTION

All too often a construction project is viewed by the parties involved as kind of a chess game; before the "game" starts, one party tries to enter the room where the chess game will take place and steal a few of the other side's pieces so that they themselves cannot lose. The other party has the same tactic: to outsmart the other side, steal a few of their pieces so that *they* cannot lose. To leave this parable: the tactics are "smart" procurement on the side of the Employer (hereinafter referred to as "Owner" except where the text refers to the "Employer's Requirements"), so that the project will be executed within the budget by not disclosing certain issues that may cause a significant raise in the price and time needed for completion. Then on the other side "smart" tendering by the Contractor where he is the counter party of the Owner in the contract, and he perhaps has discovered (some of) those issues in the invitation to tender that offer him a wide-open chance for claims during the execution. In its worst form: tender at 80% of the cost and claim an extra 40% and still make a good profit.

But construction is not a chess game; if it is a game at all it should be one with two winners, not one. This book is all about how to achieve that, two winners, the Owner and the Contractor. Actually, there should be *three* winners: the project itself should also be a winner, the primary winner; if it is not, one of the others, or both, will not be winners.

This book is not intended to be a legal guide but as a practical guide for those involved in construction "hands on" in some form of management role, be it company management or management roles in the realisation of projects. Hence the title of the book.

From the moment the idea of a project is conceived by someone, it all begins: "What do we need and who do we need to realise this project?" If the Owner is the one who sees the need for a construction project, he must evaluate from that very moment who are needed on his side to realise the project: his own staff; sometimes it may be outside consultants; maybe some public authorities will have to be involved; and other stakeholders, who will be influenced by the project either during construction or thereafter or both. It is of vital importance to get all involved "on the same page" to align the visions and the necessary actions, to even begin to make the project possible. To unify all on one's side is essential, and this should be done by focussing on what is necessary to make the project a success: "The achievement of all project objectives" as Arent van Wassenaer calls it in his excellent book *A Practical Guide to Successful Construction Projects* (van Wassenaer, 2017)[1]. The objectives for a successful project he describes are:

- Completion and operation within the allotted time;
- Completion within everyone's budget – including a fair reward for the Contractors, provided they performed as expected;
- Achieving the approval of all relevant stakeholders in terms of execution and outcome;
- Performance as required; and
- Not having given rise to any disputes during construction.

One might add:

- Without any accidents that have caused death or injury.

A potential trap may be the personal ambition of the persons involved on behalf of the parties in the project as the author highlighted in his article "Correct Scoping of the Employer's Requirements" in the *International Construction Law Review*[2]. Someone may see a career opportunity by "cleverly procuring a project for a low price" or a "clever" tender for a project by low pricing of the work and seeing opportunities to claim extra cost because of issues not addressed in the invitation to tender and/or Employer's Requirements, but that are necessary for completion of the project.

The objective of the parties involved in a Turnkey project is, or should be, to complete the project successfully, and the challenge to both parties is to realize that objective. It is a challenge because Turnkey projects are generally complicated. The complexity can be compensated by creating a strong foundation for the execution of the project by the Owner clearly and unequivocally stating what it is that he wants in the "Employer's Requirements" as to the purpose of the completed Works, functional specifications, scope of work, timeframe for construction and technical specifications, if any.

Formulation of the Owner's Requirements, in construction contracts mostly called Employer's Requirements ("ERs"), is a complicated and sometimes daunting task. The ER's and data supplied with them may leave "white spots": missing information that however is indispensable. After all, a significant majority of disputes on major construction projects find their origin in 1. issues relating to reliability of information or lack thereof and 2. deficiencies and/or ambiguity in the scope of work.

(Engineering) Contractors design and build facilities for Owners, public or private. These projects may include the design of industrial processes for these

1 Arent van Wassenaer: *A Practical Guide to Successful Construction Projects* (Informa Law from Routledge, London, 2017).

2 Eric Eggink: "Correct Scoping of the Employer's Requirements". *International Construction Law Review,* 2017, Part 1, pp. 4–15.

facilities. Some may offer Project Management for the realisation of these installations through Engineering, Procurement and Construction Management ("EPCM") contracts. Then other contractors do the actual construction under a separate contract between them and the Owner. Contractors may offer also to cover the entire realisation process from conception to delivery and sometimes even beyond, including financing, operation and maintenance (DBFOM).

For the realisation of all this, Contractors conclude contracts with the Owners, sometimes "bespoke" contracts, in the Owner's format (predominantly in the Oil and Gas Industry) and sometimes based on standard formats FIDIC, NEC 4 (mostly for Owners in the public realm), or Crine Logic (Offshore and Subsea Oil & Gas, or ICC Model Turnkey Contract for Major Projects). In all of these standard forms it is written down what is to be done in a specific project, what will be the financial compensation for that effort, how to establish what exactly was performed by testing and what would happen if the agreed performance was wrong or absent all together.

Contractors wish to receive fair payment for their efforts, and the Owner wishes to pay a fair price for receiving what he bargained for. Frequently the ideas of what is "fair" differ from one another. Contracts should cover the various phases of a project. Each project has different phases; the Contractor may be involved in some or in all of them, depending on how much scope of work he is willing to take on.

This book intends to take the reader through those phases and show what the specific risks are in a particular phase and how to mitigate such risks and how to achieve a good project in the end. These phases are:

1 Feasibility study,
2 Pre-FEED and/or a FEED,
3 Formulation of the Employer's Requirements,
4 Design,
5 Construction, and
6 Testing and hand-over.

When an Owner wants to determine if a certain new facility that he wants is feasible, or whether any optimisation is economically viable, he will do a feasibility study, or ask an external consultant to perform such a study, to see if it can be economically sound to do such an investment and to determine the scope of such an investment. Can the product output be marketed, etc.?

The study will result in a study report, and the Owner subsequently will base his decision on this report. In the next phase, there is the Pre-FEED and/or the FEED (Front End Engineering and Design) which lists the necessary engineering services, hardware and construction Works, all including price estimates.

Thereafter the Owner will generate the Employer's Requirements comprising:

1 The definitions of the defined terms in the contract;
2 When the contract enters into force;

3 That the parties must act in accordance with the principles of Good Faith and Fair Dealing towards each other;
4 The ruling language of the contract and communications;
5 Applicable substantive law;
6 New or changed laws, standards and regulations, what to do if they affect the performance of the contract;
7 Interpretation of the contract, order of priority of contract documents;
8 Gaps in the documentation and information, what to do, who is responsible;
9 Obtaining permits, planning and other permissions;
10 Guarantees, bonds or other securities;
11 Demands under guarantees, bonds or other securities;
12 No agency relationship, parties independent from one another;
13 Contractor's general obligations;
14 The Owner's general obligations;
15 Coordination with other Contractors on site by the Owner;
16 Quality assurance;
17 Staff and labour;
18 Representatives of the parties;
19 Each party's duty to notify;
20 Scope of the contract price and Works;
21 Setting out on the site;
22 Site information for the Contractor's design and use;
23 Unexpected artificial or physical conditions or obstructions;
24 Health and Safety Plan requirements;
25 Public convenience;
26 Environmental protection;
27 Services or supplies to be supplied by the Owner or public utilities;
28 Owner-supplied plant, materials or equipment;
29 Responsibility for shipping and storing goods and materials;
30 Ownership of goods, materials, plant and equipment;
31 Keeping the site clear;
32 General allocation of responsibility of the parties for the design;
33 Design review process;
34 Intellectual property rights;
35 Variations and changes to the Works;
36 Start date, Time to Taking-Over, scheduling and progress;
37 Delay damages;
38 Schedule of contractual dates, time schedule and reporting progress;
39 Progress reports;
40 Contract price;
41 Payment of the contract price;
42 Applications and procedure for payment;
43 VAT/GST or equivalent consumption taxes;

44 Financing charges for late recognition of rights between the parties;
45 Completion;
46 Commissioning and performance tests and retest;
47 Taking-Over/provisional acceptance;
48 Assessment procedures after Taking-Over;
49 Defect Correction Period;
50 General principles of risk allocation and responsibility;
51 Liabilities of the parties and their reciprocal limitations and exclusions;
52 Fraud, wilful misconduct or gross negligence;
53 Warranty for future purchasers or Operators;
54 *Force Majeure*;
55 Suspension, termination and/or rescission of the contract;
56 Insurance;
57 Confidentiality;
58 Prohibition of corruption and compliance with applicable laws and standards of ethics;
59 Exclusive remedies, entire agreement, severability, amendments to the contract, waivers;
60 Joint and several liability;
61 Assignment of the contract;
62 Communications, including notices and consents;
63 Claim and dispute resolution procedures;
64 Arbitration.

If there is a process design for a chemical process involved, the next phase will be a Basic Design and Engineering Package containing the process design executed by the Contractor. Thereafter via Detailed Engineering to Procurement (Engineering + Procurement). The next phase would then be with construction management services (EPCM), or in its most extended form: Engineering, Procurement and Construction ("EPC") and Lump Sum Turn Key ("LSTK"), all under the Contractor's responsibility in accordance with the contract. These steps are further clarified later.

The construction industry knows various standard forms of contract, such as the ICC Model Turnkey Contract, the NEC 4 Contract, the FIDIC forms of contract, the Crine Logic forms of contract, etc. They give a good idea of what is standard usage of trade in the industry and in adjacent industries.

The best contract for a project is a balanced contract

In the end, it is all about the Contractor building what the Owner wants; it is a project that both parties want to see succeed and it should not turn into a battle because of unbalanced contractual positions, causing unrealistic and unfair risk allocation that may end up in World War III. As no contract execution process will ever be perfectly smooth, certainly not for both sides, the prevention of and/or quick resolution of disputes is of paramount importance to the progress and well-being of the project itself[3].

The IACCM (the International Association for Contract & Commercial Management) through research has identified nine principles that underpin successful outcomes: 1. communication; 2. risk allocation; 3. problem solving; 4. no-blame culture; 5. joint working; 6. gain and pain sharing; 7. mutual objectives; 8. performance measurement; 9. continuous improvement[4].

If disputes arise, swift and early resolution is the key. One needs to consider if it is worth it to invest in prevention of disputes by having a Standing Dispute Board (a group of usually three people from outside the parties) appointed at the start of the project and that is tasked to resolve disputes that arise during the project. Dispute Boards are not (yet) very common in the oil and gas industry, whilst they are beginning to be used in e.g. the heavy civil construction world and have proven their value in dispute resolution. A logical step would be to expand their use to the prevention of disputes rather than the resolution of disputes, as disputes are disruptive to the process and thus to the detriment of project execution, and the sooner developing disputes are resolved, generally the better that is for the project's results. This can be done by appointing what is called a "Standing Dispute Board" following contract signature which will allow the Dispute Board to follow the development and progress of the project's works and to settle issues before they become disputes.

This should not come as a surprise; one would say that everyone involved in the construction industry – be it in heavy civil engineering or oil and gas engineering or any other kind for that matter – knows this in their heart but people seem to lack the trust in each other and the belief that one could actually relate to the other in that way without being "taken for a ride" by the other party.

Let us line up the main causes of disputes in construction projects and see how a project starts from the very beginning and how it should develop, what each of the parties should do and refrain from doing to make the project a success and what major issues have the potential to develop into disputes during the course of a project. As will become evident in Chapter 12, the Contract Manager plays a very significant central role in achieving project success just as the Project Manager does and, let's face it, as well as everyone working on the project.

The reader will discover this in the next chapters.

3 Arent van Wassenaer: *A Practical Guide to Successful Construction Projects* (Informa Law from Routledge, London, 2017).

4 See: www.IACCM.com

CHAPTER 1

Main causes of disputes in construction projects

1.1 Introduction

This chapter lists the various problematic issues that often arise in construction and explains why they develop into disputes and what can be done to avoid them.

Can one identify common categories of causes for disputes in projects, whether the project is the formulation of a feasibility study, a Basic Design and Engineering Package or an EPC/Lump Sum Turn Key (LSTK) Project, and if so what are these categories?

The answer to the question "Can one identify . . . ?" is: "Yes one can!" There are categories of causes identifiable and most of these categories apply to any type of project, whether it is a study or an LSTK project.

Problems, possibly developing into disputes, often originate in relation to:

- (Supply of) information;
- Scope of work;
- (Lack of) access to the site;
- Clash of personalities in Project Management on the Contractor's and on the Owner's sides;
- Price;
- Payment;
- Risks originating outside of the parties' control; e.g. unforeseeable physical conditions, *Force Majeure*, etc.;
- Delay;
- Quality (Performance);
- Damage and/or loss; injury or death;
- Flexibility in dispute resolution.

The vast majority of problems (if not all) between the Owner and the Contractor in a project can be related to these categories and when drafting or negotiating a contract or managing a contract during the course of a project, one should always be aware of this and seek a solution to a problem that in the early stages is perceived as a potential problem during the drafting and negotiation phase, or when it has arisen during the course of a project.

The reader should invest time in reading Professor Bent Flyvbjerg's[1] works on why cost-overruns and delays keep on happening in projects, particularly in the early phase of a project. Many think it is because of inaccuracy: unreliable data and outdated data, an unsatisfactory explanation as numerous projects have been executed, yet no improvement has been made over time where one would reasonably expect improvement.

Flyvbjerg sees "bias" as the culprit:

1. Psychological bias; optimism which is in fact self-deception: believing facts are in reality better than seen in investigations;
2. Political bias; strategic misrepresentation by overstating/overestimating benefits and understating/underestimating costs.

As an example: the author was present as Contract Manager and legal advisor in a Steering Committee meeting when the Joint Ventures of Dutch Contractors had handed in their bids for phase one, the civil works, of the High-Speed Rail Link from Amsterdam to Brussels. Total of the lowest bids was 5.2 billion NLG (or today: €2.36 billion). During that meeting the news broke that the minister of transport had just addressed the Dutch parliament stating that she thought the whole line could be built for 3.8 billion NLG (or today: €1.7 billion). The predominant question was: "Why would she say that when the bids had just come in?" That became clear soon thereafter, when the minister announced that the government would procure the "Betuwe Route", a freight rail line from Rotterdam to the German border: she needed parliament to allow her a budget for that project. Even until today on the German side of the border the line has not yet been connected further into Germany. But the minister very much wanted to portray a "green image".

1.2 Supply of information

Particularly in the early stages of a project the Contractor is depending on information given to him by the Owner either during the bidding phase or during the early phases of the project. How far can the Contractor rely on the correctness and completeness of the information received from the Owner?

A lot of information may prove to be crucial for the success of the project, and the Contractor must be able to rely on the facts in the information and the completeness of the information, or, if that is not the case, he must be allowed to investigate the reliability of the information. E.g. correctness of subsoil information will be vital for the design of the foundations and the stability of the foundations is crucial to the safe erection and operation of the rest of the project.

In the event of a "revamp", a restauration and improvement of an existing facility, it will be the "as-built drawings" of the facility as it was built, if any. If the Owner indicates that information can be relied upon – generally this is

1 Bent Flyvbjerg: *Mega Projects and Risks* (Cambridge University Press, Cambridge, 2003.

done by marking the information as being "rely-upon", the Contractor will treat it as "rely-upon information" and not conduct any further investigation as to its reliability.

It is prudent contract management to list all the information that is indicated as being "rely-upon" in a separate document, and it is even better to attach such a document as an attachment, signed by both parties, to the contract. If the rely-upon information turns out to be wrong the consequences thereof will be for the risk and account of the Owner as the issuer of the (rely-upon) information. Information that is wrong or flawed will cause change orders, resulting in increased cost and delays in the schedule for performing the project, all at the risk of the Owner, as the Owner issued the incorrect information.

One might ask: "Then why would an Owner ever issue rely-upon information and set himself up?"

But one needs to realize that not issuing such information increases the risk to the Contractor significantly and as a consequence the price will increase significantly as well as the time for completion the Contractor will think he needs. The project will suffer substantially. Thus, it is of paramount importance that the Owner prepares the information with due diligence; it will form the very foundation of the project.

Perhaps another answer would be to give the bidding Contractor sufficient budget to do the necessary investigations, but to give all bidding Contractors a budget to do so (in case of a bidding competition) could become very costly. To allow tenderers to share in the execution of one investigation increases the risk of collusion as they might talk to one another about other things than just the resulting information.

There is the same risk of having negative cost and negative time effects if such "rely-upon" information is given later, during the course of the project, as such information may require changes in the engineering and changes in the equipment and/or materials to be delivered, or in completed sections and thus have an impact on the schedule as well as the cost of the project. If the Owner is not sure of the correctness of the information he should say so and the Contractor should then say that further information and thus investigation is necessary, reasonably to be paid for by the Owner. Information may also come without any warranty from the Owner, and in that case, the Contractor may not be able to rely upon such information other than at his own risk.

Therefore, a contractual arrangement for "rely-upon information" is strictly necessary. A provision in the contract may be worded along the following lines:

> Owner shall indicate on information provided to Contractor when such information may be relied upon by Contractor. Contractor may then rely upon such information and where any incorrectness in such information causes the need for a change order the Contractor shall then be entitled to compensation of reasonable cost and profit as well as to additional time for completion.
>
> In such event, the Contractor must act in accordance with the "variation" provision of the Contract.

The Contractor logically is always responsible for the interpretation of, and conclusions he draws from, the information provided to him.

Often the Owner knows exactly what he wants and provides the Contractor with specifications ("specs") of exactly what it is he wants. Based on those specifications the Contractor does the pricing of and the bidding for the project, the process design/Basic Design, etc. Sometimes Owners do not specify in detail what they want other than the desired performance of the Works and leave it to the Contractor and the Contractor acts on his own technical specifications which will lead to the desired result, expressed in the Owners functional specifications. This happens regularly with regard to oil and gas installations.

As the Contractor's own specifications often tend to end up in lower overall cost, the Contractor sometimes bids on his own specifications rather than on the Owner's specifications in order to be able to present a lower price and be invited to negotiate for the contract. This can be a very risky strategy for such bid may (actually: should) be ignored as it does not comply with the Owner's specified preferences. If the Owner is tempted to accept the bid purely on the basis of the low price, he runs the risk of not getting what he wants but "getting what he paid for". It is not uncommon and actually might be sensible for the Owner to reject an abnormally low bid, a bid that is substantially lower that the other bids received. Accepting such a bid may mean that the Owner will be hiring either the dumbest or the most desperate Contractor, which obviously will not lead to a successful project. Some Contractors follow that risky bidding strategy because they need work and cash flow to be able to pay wages, debts, etc.

It is of paramount importance that the supply of information happens when it is needed as to not hamper the design process or cause the need for design changes because the information was provided too late in the progress of the design. What will cost €10 to change on the drawing board may cost millions to change during construction. Information has a direct effect on the following issue: the scope of work.

1.3 Scope of work

A project starts with a Design Brief, a document generated by or on behalf of the Owner which formulates the Basis of Design. The Design Brief in itself is generally a multi-paged (easily 40+ pages) document that specifies in detail (functional and otherwise, e.g. technical) what the Owner requires and what the *purpose* of the Works will be after completion: what the Owner intends to use it for and how he intends to use it and thus what the Contractor must produce. From this flows the extent of what exactly the Contractor is to do and to produce, that extent is the scope of work usually laid down in an appendix to the contract. The scope of work and the functional specifications therein comprise *what* the Contractor must produce; the technical specifications comprise *how* the Contractor must produce, and in a Turnkey project the Owner should avoid technical specifications and rely on the Contractor's know-how and not subject himself to the risks and liabilities of defects in the technical specifications.

The trouble with formulating the scope of work is that it is rather difficult to formulate exactly what the Owner wants and what the Contractor is to produce. Words are often open to interpretation and therefore much less descriptive than diagrams and numbers.

For each of the project phases, from the Basic Design and Engineering Package (BDEP) through start-up, there is a scope of work, which includes the specifications, which define what the Contractor is contractually bound to perform and produce. In practice, all too often scope of work items may be scattered over several appendices.

It is a primary task of the Contractor's proposal team to know what the full scope of work is that is to be priced. For the Contractor's management, it never hurts to ask if the proposal team has perused all the appendices of the Owner-supplied documents for scope items: "Better safe than sorry!"

Where there are unclear issues, the Contractor must ask for clarifications in the bidding phase and he must resist the urge not to ask these questions fearing that all his bidding competitors will hear the answers as well. That may be so, but by asking for clarifications the chances (read: risks) of making mistakes that may have significant budget and time consequences will be greatly reduced. Also, what would the Dispute Review Board or the arbitrators decide if such obviously necessary questions were not asked?

Based on the specifications, the Contractor quotes his price and the Owner eventually, after negotiations, comes to an agreement with the Contractor on the contract price. Therefore, if there is any substantial change to the scope of work, such change will most likely have an impact on the cost (and the schedule) and in that case must have an impact on the contract price and the time for completion.

One can already imagine that under a fixed price contract, any changes to the scope of work are a source of debate between the Owner and the Contractor, whether or not such a change is already within the scope of work or a within a variation to the scope of work and therefore if the Contractor is entitled to additional payment (and/or time for completion, as the case may be) – or that it is not achievable.

The more detailed the technical and functional specifications are formulated the more the Owner is responsible for the choices he has made. It is the Contractor's duty to check such specifications against the functional specifications of the end product expected by the Owner and to see if the end product is achievable with the technical and functional specifications as given by the Owner, and to notify the Owner if he is of the opinion that it is not.

For the Owner, there is an eternal tension between his available budget and the total cost of the facility he is procuring (see earlier: Flyvbjerg's opinion on "bias"). Often Owners mistakenly believe that "the more you specify, the higher the cost will be" and this belief will lead to condensed, general and thus less descriptive specifications and thus a diffuse scope of work.

For the Contractor, certainly when he is expected to quote a lump sum price and accept the full price risk, unclear specifications form a high-risk factor: he quotes a fixed price while he is not sure what exactly he is to deliver. His room for including contingencies in his price is very limited; he is bidding in competition with others and adding contingencies may mean his price will be too high to be selected as the preferred bidder. For the Owner the result will be that he most likely will award the contract to either the dumbest or the most reckless and desperate-for-work bidder.

One can easily understand that this can lead to severe conflicts during the execution of the project: while the Owner will be of the opinion that an item is included in the price the Contractor will be of the opinion, if it is not expressly specified, that it is not included in the price. See further under Section 1.5.

Scope of work may also be fluid in the sense that it will be impossible to establish the exact scope of work because an existing facility is to be restored, refurbished or revamped without there being accurate "as built" drawings available. Let's take a 600-year-old cathedral as an example. With every section of a wall, the question will be: "What will we find behind it?" In such a case it will be impossible for the Contractor to quote a fixed price other than for the known scope of work and he will have to exclude issues that are uncertain.

A non-limitative list of such exclusions in general:

- Discrepancies between information and data provided by the Owner and "as-built" status (e.g. errors, omissions or inaccuracy, deterioration of equipment, materials and current performance) occurring between contract award and start of the Works;
- Hidden defects and/or hidden conditions of existing equipment and materials (including but not limited to non-weldability of existing/aged materials, fouling);
- Presence of physical conditions, hazardous materials or substances (including but not limited to asbestos, contaminated soil) or the occurrence of underground obstacles and removal/disposal thereof;
- Deterioration of existing civil foundations and their quality not known prior to contract award;
- Consequences of deviations in design feedstock characteristics;
- Impact of HAZOP (Hazard and Operability study);
- Impact of local authority requirements on the non-modified parts of an existing facility (the parts that will not be changed and will be left as they are) to comply with the latest regulations with regards to emissions, effluents, noise, modified flare load and its compliance to authority rules, etc.;
- Changes in laws, regulations of the country, including tax laws or tax regulations, relating to design, environment, health and safety, introduced after the Effective Date of the contract;
- Cost of rectification of loss or damage caused by a *Force Majeure* event;
- Demurrage costs related to construction not attributable to Contractor.

1.4 Access to site

Once the project is in execution it is vital for the Contractor to have physical access to the site. It is not enough that the site is formally handed over to the Contractor but he must in fact be able to get to the site and get onto the site.

Example: a Contractor was to build two Ferry Terminals on either side of a river in the middle of a tropical jungle.

The access road was 30 km long and had been assessed as being an acceptable yet dirt road during a site visit in the tender phase by the interested Contractors. This visit had taken place in the dry season, but when the project execution on site was to start, the rainy season had started and consistent excessive rain had rendered the road completely useless to transport building materials and equipment to the site.

The quartermasters had looked for alternative ways to get the materials and equipment to the site and had found the river itself an adequate way of transporting materials and equipment to the site, that is, until they got shot at by smugglers who used the river to transport their contraband and would not tolerate any “peeping toms”.

So, the only alternative was now useless as well. When the quartermasters were asked by management if they had at least sent a written notice to the Owner that the access road was unusable and therefore the Contractor had no access to the site, their answer was: Who starts a project by sending such a notice to the Owner and starts the project by “planting a knife in the table”? In other words, they had not sent a notice. The basic sentiment is understandable: Is it sensible to immediately start sending notices to the other party if something is not quite right; will that not ruin the good atmosphere on the project that is necessary to make the project go well? But when because of something not being right there are major time consequences, it would be a very costly mistake not to send such a notice, particularly under a fixed time contract. Besides, it might be beneficial to the Owner to know about trouble brewing when trouble announces itself.

It can also be that at the time of commencement of, or during, the Works at the site, the site is flooded and therefore inaccessible. If a Contractor cannot access the site due to any reason not caused by himself, the prudent course of action is to send the Owner a notice to that fact. On a project the Owner was to keep the site clear of flood water. The site flooded and the Contractor sent timely notice of the fact and claimed additional time equal to the time the site would be flooded. The Owner answered that the contract stated that the Contractor was to foresee all risks and thus would not be granted an extension of time, a nonsensical reply as the Contractor does not have to foresee that the Owner would not honour his contractual obligations and be in breach of contract.

In addition to that: most contracts limit the time for giving notice from the moment any event gives rise to extra cost and/or extra time and as soon as the Contractor knew, became aware of or should have been aware of such an event, he must give notice of such event. The Contractor often has a contractually limited number of days for giving such notice and exceeding such number of days may very well lead to forfeiting one's claim to extra time and/or extra money depending on the contract conditions. The Contract Manager should have an A3 size paper hanging on the wall of his office displaying just the number of days one has to notify the other party of anything that must be notified. In troubled projects the discussion will focus on when the Contractor should have notified the Owner of any issue discovered that may have a cost and/or time impact.

1.5 Clash of personalities

It is not uncommon that irritations quickly become predominant because the Project Manager of the Contractor does not get along well with the Project Manager of the Owner or vice versa. Sometimes people rub each other the wrong way and for some reason or another they cannot step over these issues in the interest of the project. It is hard to determine where (or: in whom) the real cause of trouble lies, but there is no denying that there is trouble sometimes.

As an example: the Contractor sends notices to the Owner and the Owner's Project Manager sees this as claim-oriented behaviour and foresees that during the entire time of the contract execution this will be a struggle for him, thereby overlooking the fact that such notices may in fact only be intended as warnings that not everything is 100% on the Owner's side and the notice is merely intended to prevent things from getting worse as they progress while they are under the responsibility of the Owner.

It is equally possible that on the Contractor's side not all is 100% in order, and why would one object to a warning that such issue may have an effect either in time or in cost or both when he is still in the position to do something about it and fully mitigate those effects, e.g. a very unorganised site that slows the work down?

All too often, we choose our perception and in doing so, we may make the wrong choice. Leading in our choices should be the benefit of the project and achieving an excellent final result, and one should avoid letting one's ego get in the way of that. It is not always as simple and obvious as it may seem.

As a leader, be it CEO of an Owner or CEO of a Contractor, a Project Manager, etc. one must do what is necessary for the people in one's company or team, to be able to do what they are good at and to get even better at what they do. For both the Owner and the Contractor that means selecting a person who is a good leader who can deliver a good project, not someone who is just tough and will be able to push his own agenda through, unless that agenda is to achieve a good project which answers to Arent van Wassenaer's mantra[2]:

- Completed and operated within the allotted time;
- Completed within everyone's budget – including a fair reward for Contractors provided they perform as expected;

2 Arent van Wassenaer: *A Practical Guide to Successful Construction Projects* (Informa Law from Routledge, London, 2017).

- Meets the approval of all relevant stakeholders both in terms of execution and outcome;
- Performs as required;
- Has not given rise to any disputes during construction.

1.6 Price

Pricing a contract has many options varying from

1 A complete lump sum for all activities (scope) under the contract to
2 Completely reimbursable whereby each activity and each item to be delivered has its own rate and is recorded and whereby the final price of the contract is only established at the very end when all activities and deliveries have been completed and measured and multiplied by their specific individual rates.

There are many hybrid and mixed forms of pricing a contract and they are always in between lump sum and reimbursable.

The Contractor will be prepared to quote a lump sum price when he feels secure about what activities and deliveries are necessary; the more routine the more secure. The fewer uncertainties in the scope of work, the clearer the pricing becomes and the easier it is to quote a lump sum, also known as a fixed price. The more uncertainties in the scope of work, the less certain the pricing becomes; a Contractor would only be prepared to undertake the work on a reimbursable basis, quoting rates for man-hours and material to be used.

A classic example of this is the earlier mentioned restoration of old buildings, like cathedrals, etc. One never knows what is hidden behind what one can see. In oil and gas engineering a classic example can be a revamp (renovation) of an existing refinery installation.

The Owner's preference will generally be for a lump sum, as that makes it easier for him to budget the project, but some Owners may think that certain projects should be on a re-measurable basis if this allows for better control of the expenditure on that project. But that also means that the Owner will need an extensive staff able to monitor all the activities of the Contractor under the contract. Contractors will not be inclined to give the Owner the right to audit the components of a lump sum price, whereas the reimbursable components of a price must be auditable by the Owner.

Advisory services during construction, as well as during pre-commissioning and start-up cannot be priced on a lump sum basis, as one cannot be sure when these services are to be provided and for how long.

In a lump sum, or fixed price contract, if – during the course of the project – the scope of work changes, the price will also have to change. This is done via a "change order" or "variation order" stating what the change to the scope of work exactly is and what the price consequences and also the time consequences of that change order are. Later in Chapter 9 there will be extensive attention to change orders and their consequences.

Also, during the course of a project, the local laws or regulations may change, particularly within the realm of taxes, customs regulations, building codes, environmental impact, etc.

These will be mandatory laws and regulations, and the legal obligation to follow these laws and regulations may very well have an impact on the cost and scheduling of the project and must therefore also have an impact on the price of the project and its performance schedule. In other words, there must be a provision in the contract to adapt the price for the consequences of changes in laws and/or regulations if such changes could not be foreseen during the bidding stage.

An important issue regarding price is the currency in which the price is expressed. Often, but certainly not always, a Contractor expresses his price in the currency of his home country. As soon as he quotes his price in a currency other than say his "home country currency", the risk of the exchange rate between his home country currency and the quoted currency is yet another threat to the Contractor's earnings on the project.

To give an example: if a Contractor quotes €100 million as a price for a certain project, he knows that if everything goes as planned he will receive a total of €100 million over the course of the project.

If he has calculated his price to be €100 million but has to quote in US dollars then the problem is that the exchange rate of the Euro against the US dollar changes almost every day. So, when he earns dollars, gets paid and has to change the dollars back into Euros, the amount in Euros will depend on the exchange rate at the date of the swap from dollars to Euros. Say the Contractor quotes the price for the project to be US$120 million as at the moment he quotes the price to the Owner the exchange rate is US$1.20 for €1. Now say the Contractor receives a 10% Advance Payment, thus US$12 million. Assume that at the time of the payment of the Advance Payment the exchange rate between USD and EUR is 1.10; thus, instead of receiving €10 million he only receives €9,090,909, so he loses roughly €901,000 on this one transaction alone. Things could be even worse if the exchange rate changes for the worse. Admittedly, it could also get better but a sensible Contractor will not want to gamble with his own money. So, he needs to make sure he gets €1 for every US$1.20 and the instrument to make sure that will happen is called: "Hedging". Two common ways of hedging are forward contracts and option contracts.

A forward contract will set an exchange rate for the date on which the currency transaction will occur in the future. With this forward contract the company forfeits the gains if the currency fluctuation would be in its favour, but it protects the company from fluctuations to its disadvantage.

An option contract sets an exchange rate at which the company may choose to exchange currencies. If the current exchange rate is more favourable, then the company will not exercise this option.

The main difference between the hedging methods is who derives the benefit of a favourable movement in the exchange rate. With a forward contract, the Owner gets the benefit while with an option contract the Contractor retains the benefit by choosing not to exercise the option if the exchange rate moves in his favour.

1.7 Payment

Payment issues may arise due to various causes. Potentially the first payment issue is with the advance payment the Owner has to pay to the Contractor to cover the Contractor's initial cost at the start-up of the project. Such advance payment is a risk to the Owner, as he must pay before having received anything. To cover this risk Owners usually want to receive an on-demand advance payment guarantee from the Contractor's bank. The Contractor's bank will be willing to provide such a guarantee if the Contractor has a sufficient financial position with that bank to cover the obligation of the bank to pay on the advance payment guarantee.

Usually the advance payment is paid back by the Contractor to the Owner through deductions on the payment by the Owner on invoices of the Contractor. If the advance payment was say 10% of the contract price, then on every invoice issued by the Contractor the Owner pays 90% of the value of work and materials on that invoice and thus 10% of the invoiced amount is deducted in the invoice until such time that the entire advance payment is deducted and thus repaid.

The advance payment is intended to prevent the Contractor commencing the project with a substantial negative cash flow and thus a risk exposure that the Contractor has no control over.

If the advance payment is to be repaid through deductions from the amounts payable on subsequent invoices during the course of the project, then the advance payment bank guarantee must reflect that option in the text so that the guaranteed amount decreases with the amount of every repayment. If it does not reflect that, the exposure of the full amount of the advance payment remains through the outstanding bank guarantee while the advance payment is being repaid.

More prominent are the problems caused by late payments by the Owner. This is a problem that does not exist if the Contractor receives a letter of credit for the contract price minus the advance payment from the Owner and where the Contractor can draw money from the letter of credit with approved invoices. As not every Owner can, or wants to, have its bank issue a letter of credit, we need to have a look at the payment risk in case of a cash contract.

Late payments basically mean that the Contractor has to pre-finance the project without a contractual arrangement thereto and late payments are very detrimental to maintaining a neutral or slightly positive cash flow for the Contractor on the project.

To mitigate the risk of late payments the Contractor has two options that are not mutually exclusive:

1 A contractual interest on amounts that are overdue, e.g. the current percentage of three months Libor or Euribor (depending on the currency of the contract) plus say 5%. The idea is to make it more expensive not to pay the Contractor than it would be for the Owner to secure a loan from a bank to enable him to pay the Contractor, if the Owner has cash deficiencies.

2 To secure the contractual right to suspend and eventually to terminate all work under the contract until such time that payment, for work done and invoiced, is received by the Contractor.

Payment after all is one of the primary obligations of the Owner under the contract and defaulting on the payment obligation is a material breach of the contract, and the Contractor must have the right to suspend and subsequently terminate the contract for such material breach.

The Contractor must be able to suspend and ultimately terminate the contract if the Owner is habitually derelict in the payment of invoices. Whether or not the Contractor makes use of such a right is a matter to decide for the Contractor's management.

If the Owner does not pay due to the fact that the Contractor's invoice is not correct, the Contractor has no other option than to correct the invoice. If the Owner disputes an item of the invoice he generally is obliged to pay the undisputed part of the invoice, depending on the law governing the contract.

1.7.1 Letter of credit

A letter of credit ("L/C") is a document issued by a financial institution, usually a bank or a similar party, assuring payment to a Contractor, provided certain documents have been presented to the bank, documents indicated in the body of the letter of credit that prove that the seller has performed the duties under an underlying contract (e.g. an EPC contract) and the goods and/or services have been supplied as agreed. In return for these documents, the beneficiary may draw payment from the financial institution that issued the letter of credit. The letter of credit serves as a guarantee to the Contractor that he will be paid and thus the L/C must be irrevocable. For a Contractor based in a different country than the Owner it is advisable to get the letter of credit as issued by the Owner's bank to be confirmed by the Contractor's own bank. Such confirmation will attract a bank fee which should be for the account of the Contractor.

When the L/C is for a total of the contract price minus the advance payment the Contractor has the comfort of knowing that the money is there to pay the Contractor when he has fulfilled his contractual obligations. To ensure cash flow the Contractor usually may draw from the L/C at given intervals when certain milestones have been reached or a certain progress has been achieved. Sometimes an L/C is only for a part of the contract price and then the Owner must either replenish the L/C after a drawdown under the contract or revert to cash payments once the L/C has been depleted. An L/C that needs to be replenished is usually referred to as a "revolving letter of credit". The downside for the Contractor of such a form of payment is of course that the Contractor has no security that the Owner has the financial means to pay the Contractor, just like cash payments, well maybe slightly more security because if the L/C is not replenished after a drawdown that is a signal that the Owner may not have sufficient funds to meet his primary obligation under the contract namely to pay the Contractor for work done.

The issuing bank's role in all this is to check if all the necessary documents have been provided by the Contractor and if they conform to the bank's requirements, the bank must release the payment. The bank does not check the quality of the Contractor's performance. The Owner will have to give comfort to its bank by either putting up the cash up front, usually by reserving the equivalent of the value of the L/C from its balance at the bank, or by having a credit line with that bank.

To have the security that a foreign bank is a solvent organisation that can pay the Contractor what is due to him, a Contractor can ask his own bank, or a corresponding bank of his own bank, to confirm the L/C. As said this will attract a fee, but that is a small price to pay for the security that one will be paid for one's efforts. This is called a "confirmed letter of credit". For the letter of credit to be confirmed it must be irrevocable. The essence is that one can draw (= get payment) under a letter of credit at the bank of one's choice that has confirmed it and avoid issues of default of the issuing bank (Figure 1.1).

Letters of credit are used primarily in international trade for transactions between a supplier in one country and an Owner in another. Most letters of credit are governed by rules promulgated by the International Chamber of Commerce known as Uniform Customs and Practice for Documentary Credits (UCP 600) as an example, usually indicated in a contract by the words: the "current version" of the UCP at the time of the Contract becoming effective.

1.7.2 Guarantees

A few words on guarantees.

Owners generally do not want to pay before they have received something, and neither do Contractors want to pay subcontractors and suppliers until they

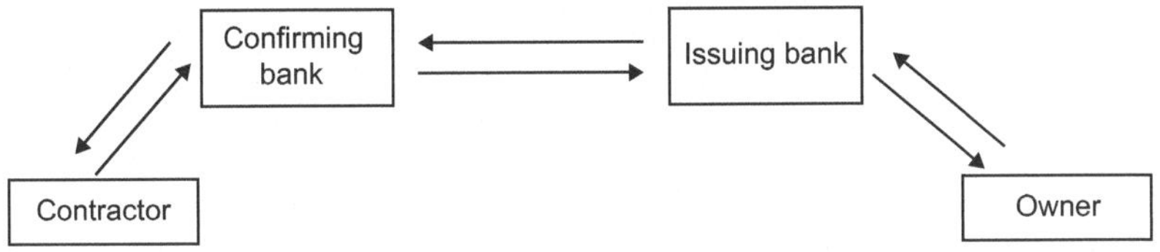

Figure 1.1 Schematically the system looks like this. The arrows pointing to the left reflect the flow of money; the arrows pointing to the right reflect the flow of documents.

have received their agreed performance. To overcome this stalemate the Contractor provides bank guarantees to secure payment from the Owner, an advance payment guarantee, a bank guarantee for commonly 10% of the contract price because the Contractor wants to receive a 10% down payment at the start of the work so that he will not have a negative cash flow. A bank guarantee is a written guarantee by a bank to pay, as a primary obligation (= direct obligation of the bank itself), a certain amount of money (10% of the contract price in our example) on the first demand by the Owner (= on-demand guarantee). It is prudent to agree to have guarantees issued under the ICC's URDG 758 rules. URDG stands for Uniform Rules for Demand Guarantees.

All the Owner has to do is write the bank a letter in which they "call" on the guarantee. This poses a risk to the Contractor as the Owner might call the guarantee without a contractual right to do so (unfair calling).

To protect oneself against unfair calling Contractors do well to agree, prior to the signing of the contract, the wording in the guarantee that requires the Owner to state in the letter, with which they demand payment under the guarantee, to state the exact (contractual) reason why the Contractor is to be considered "in default" and stating in that letter the exact reason for calling the guarantee. Unfair calling of the bond happens when a bond is called without good (= just) reason. It happens that in the Owner's organisation it is considered that when the time comes (when the Contractor challenges the calling of the guarantee) the Owner will then come up with a reason. A Contractor can also take out insurance against unfair calling, but obviously that will cost money (premium) and the worse the Owner's payment reputation is, the higher the premium will be and the cost of insurance will increase the contract price.

It is good and prudent practice that the guarantee contains wording that states an end date for the bank guarantee and that the guarantee will not be of any value past that end date, to make sure that the bank will not have to pay beyond that date and that the Contractor is relieved of the obligation to maintain the availability of either the funds or the credit line to cover the amount of the guarantee.

In giving such a bank guarantee, the bank obviously wants to make sure that they can take recourse against the Contractor by making sure that the counter value of the bank guarantee is either in the corresponding bank account of the Contractor or secured by a credit line of the Contractor with the relevant bank. This is called the *obligo*.

The same can be said for performance bank guarantees which the Owner usually requires to make sure that whatever the Contractor builds conforms to the specifications and Works as promised and will continue to do so at least during the Defects Liability Period as stated in the contract. If the performance guarantees are not met or materials fail, that entitles the Owner to call the bank guarantee. If through delays on the project the expiry date of the guarantee falls before the performance test is executed the Owner would require either an extension of the bank guarantee or payment of the guaranteed amount of money (= "extend or pay" call).

As an additional security the Owner may require a parent company guarantee, sometimes even from the ultimate parent if the Contractor is part of an incorporated group of companies. In such a guarantee the (ultimate) parent company guarantees the performance by its subsidiary, whereby the Contractor will require wording that the parent company takes on no more obligations than the subsidiary did in the relevant contract. This parent company guarantee also needs an end date and it will cost the subsidiary money as payment for the risk the parent company takes, as such a guarantee will appear as a potential payment obligation on the balance sheet of the parent company.

1.8 Risks originating from outside the contractual parties

During the course of a project a number of risks may materialize that have a significant impact on the project economics but that do not originate from one of the parties.

The first category of such risks is: unforeseeable physical conditions, either natural in origin or man-made. The relevancy of such conditions is that they increase the scope of work of the Contractor and thus increase his cost as well as the time needed for completion.

Such conditions may be in relation to rely-upon information provided by the Owner, and when they materialize they will cause an increase in scope of work and thus in time for completion and cost. If such conditions were mentioned somewhere in the documentation, then they would be unforeseen conditions rather than unforeseeable conditions, because the Contractor overlooked them when perusing the documentation received from the Owner. The Owner should remain responsible for unforeseeable conditions and their consequences must be for the risk of the Owner as they will cause an increase in the scope of work for the Contractor and the Contractor cannot price those consequences nor plan measures to mitigate their effects. Examples of such conditions are subsoil and hydrological conditions as well as extraordinary severe climatic conditions.

The key issue is here: What is the pivotal point? When are conditions such that an experienced Contractor should have been able to assess that conditions in reality might be different than they appear from the Owner's documentation?

He might interpret the provided data as deliberately kept vague by the Owner for whatever reason and if the other bidders do not have the same doubts he will most likely not get the job if he expresses his. David Kinlan works out this issue very well in his book *Adverse Physical Conditions & The Experienced Contractor*[3].

An illustration: a dredging Contractor is contracted to deepen a channel to -20 metres to allow access for oil tankers to a new refinery being built. The Owner had a soil investigation done in the existing natural channel by having boreholes

3 David Kinlan: *Adverse Physical Conditions & The Experienced Contractor* (Delft Academic Press, Delft, 2014).

drilled every 50 metres over the total length of 2,5 kilometres. While dredging the channel the dredger encounters a large submerged granite "mountain top". The granite is so hard that it cannot be removed by the dredging equipment and it will require removal by blasting with explosives. The Contractor's Project Manager immediately notifies the engineer. The project is executed under the FIDIC 4th edition Contract Conditions.

The engineer reacts to the Contractor's notification that the encountering of this indeed unforeseeable condition is the Contractor's risk as the Contractor's interpretation of the borehole data are the Contractor's risk, in particular that the Contractor interpreted the data as that no such adverse physical condition would be present and thus the removal of the granite would be for the account of the Contractor, a defendable interpretation and decision.

The Contractor, knows that the removal of the granite through blasting will be very expensive turning the project from a profitable job to generating a considerable loss and possibly a delay as well. He feels that the decision of the engineer is unfair under the given circumstances, so he argues that the removal of the granite should not be for the account of the Contractor. The issue is discussed within the tender team that had prepared the bid together with some of the Contractor's experienced Project Managers. From those discussions it becomes clear that the great number of boreholes was, for this kind of work, very unusual. Apparently, the Owner wanted to avoid adverse physical conditions that might cause delay and budget overruns on this project. The conclusion was:

> it is pure bad luck that they started on the spot where they did the first borehole, 25 meters closer or further away and they would have drilled directly into the granite and we all would have known about it.

In a discussion with the Contractor's management the Owner confirmed that this was indeed the situation they had wanted to avoid and also agreed that if the first borehole indeed had been in a different place (which was chosen under the responsibility of the Owner) the granite would have been discovered and the cost of the removal would have been for their account. The Owner agreed to pay for the removal.

Lesson: these issues are not always purely legal issues and by a multi-disciplinary approach one may find solutions to a problem that may otherwise not be found. Particularly in the dredging industry unforeseeable physical conditions that compromise the Works under the contract generally have a major effect on the project and David Kinlan is right where he states in his book that the NEC 3 approach of defining quality as having everything to do with meeting the performance guarantees that the Owner has asked for and that the Contractor has given in the listing of performance guarantees under the contract. They have to do with a certain output resulting from the use of a defined feedstock under given specifications with a defined maximum energy consumption and e.g. a maximum allowable expulsion of NOx in the exhaust gasses of the plant. Quality also has to do with the quality of the supplied hardware.

Defining physical conditions which have such small chance of occurring that it would be unreasonable to have allowed for them in the contract price as a contingency is a better way of dealing with unforeseeable physical conditions than apportioning risk to the Contractor for what an experienced Contractor could/should have foreseen and that such foreseeability by an experienced Contractor would deny him any claim for extra money and extra time as the presence of such conditions increases his scope of work.

Other risks originating from outside of the parties are events that may cause physical damage to the Works or delays in performance of the work, such as:

- War, whether declared or not, violence, invasion or acts of foreign enemies;
- Rebellion, terrorism, revolution, insurrection, military or usurped power, or civil war;
- Riot, commotion, disorder, strike or lockout by persons other than the Contractor's personnel and other employees of the Contractor and his subcontractors. As many areas in the world are volatile, this may be a very relevant risk for the Contractor, one that may be hard to insure or might actually lead to the insurance companies terminating coverage for this risk in certain areas of the world at specific times. It is a risk that may not just occur at the site where the facility is to be built, but also in the place where the Contractor chooses to have certain equipment or materials made or shipped from.
- Munitions of war, explosive materials, ionizing radiation or contamination by radio-activity, explosives, radiation or radio activity;
- Pressure waves caused by airplanes or flying objects flying at subsonic or supersonic speed;
- Natural catastrophes such as earthquake, hurricane, typhoon or volcanic activity.

All consequences of physical damage due to these risks must be for the account of the Owner where such risk materializes at the site. These risks are the Owner's entrepreneurial risks. Be aware that in the more volatile areas of the world, Owners may be looking to limit liability for the consequences of these events particular in relation to consequences of human conflict and any insurance on the Contractor's side may be voided by the insurer where the insurance policy allows that through a "war risk" clause.

These events may seem to ring a bell in relation to *Force Majeure* events, and indeed they are very similar. Earlier, we discussed who should bear the risk of physical damage to the Works as a consequence of the occurrence of such events. The Contractor's insurers on the other side may limit or even exclude coverage of damages due to such events to the Contractor's plant and materials not yet incorporated in the Works.

In relation to *Force Majeure* the discussion focuses on whether or not the parties are prevented by these outside events from performing their contractual duties (for further reference see Chapter 7).

1.9 Delay

Another major cause of disputes between Contractors and Owners is delay in the delivery of the work to be performed under the contract and more precisely the cause of such delay. All lump sum Turnkey contracts and most EPC contracts will be fixed price contracts with a fixed time for completion. As no Contractor in a major project could afford the risk to accept full payment of the entire project in one bulk payment upon successful delivery of the project, there will have to be intermediate payments during the course of the project.

Payments can best be based on progress made or on "agreed milestones", i.e. defined moments of achievement of a particular performance so that dispute regarding achievement is virtually impossible. That is why payment based on milestones is far preferred over progress payments, as progress is something that can be argued over endlessly, leaving it unclear whether or not a payment obligation exists or not, not to mention that progress will have to be measured, which requires a substantial amount of well-trained and experienced staff to monitor the progress.

When milestone payments have the preference, the Contractor will wish to align payments with his expenditures as to maintain at least a neutral cash flow and he will obviously prefer a positive cash flow. A classic mistake is to link delivery expenditure with a construction milestone, in other words as a Contractor to formulate your own payment milestone as a construction completion event, while your expenditure will be triggered by delivery of hardware: Contractor will have to pay for delivery of the hardware but before achievement of the relevant construction milestone which triggers payment of the money due to the Contractor.

This is particularly dangerous if the construction milestones are sequential and the later one is dependent on the timely completion of the earlier one(s), causing all sequential milestone payments to be late if the delay cannot be overcome.

A good way to prevent such negative cash flow for the Contractor is to link payment with proof of order commitment for goods to be delivered by the Contractor's vendors or subcontractors. The Contractor gets paid by the Owner for committing himself to a certain purchase of goods for the project (order commitment). Milestones should ideally be designed as clearly noticeable events that leave little room for discussion.

The Contractor must make sure that where he is dependent on the Owner for his progress, e.g. because of Owner approval of design work or delivery of Owner-supplied hardware or work, such performance by the Owner is subject to time limits. With regard to approval of design work, it is common to set a time limit of ten to twenty days (working days or calendar days) and the contract must then provide that if the Owner does not approve within that time limit, the design work shall then be deemed approved. Be well aware however that either approval by the Owner or deemed approval of the design does not relieve the Contractor from his contractual obligations to deliver the Works in conformity with the functional specifications in the Employer's Requirements, as we will see later.

An issue of similar importance regarding delay is the triggering of liquidated damages because of a delay in performance. Liquidated damages are a pre-agreed amount of damages that the Owner may collect as indemnification upon a specific breach of contract, usually delay but also, depending on the type of contract, the built plant not achieving the performance guarantees. See hereunder under Section 1.10 "Quality".

Liquidated damages for delay are therefore a pre-agreed indemnification by the Contractor for damages caused by a late performance usually expressed in a certain percentage of the contract price per week. Normally such liquidated damages are capped to a certain maximum percentage of the contract price as to limit the exposure of the Contractor to damages for delay. Example: 0.25% of the contract price for each full week of delay with a maximum of 5% of the contract price. It is not uncommon to have a grace period of a number of weeks preceding the triggering of liability for delay.

Liquidated damages must be the sole financial remedy for delay; there should not be other financial compensation for delay under the contract or at law, unless the delay was caused by wilful misconduct or gross negligence. The obligation to deliver the Works will however remain; simple delay should in itself not be ground for termination of the contract, unless the maximum amount of delay damages in the form of liquidated damages have been incurred. That would constitute material default under the contract. Then it is up to the Owner to decide if the contract is to be terminated.

Liquidated damages must be distinguished from penalties. Penalties are not a compensation for damages and therefore after the payment of penalties one is still liable for the damages due to the delay that is attributable to you. In some jurisdictions (e.g. England and Wales) penalties are not allowed while in others they are, so it is advisable that the Contractor seeks legal advice on this issue before making his bid.

1.10 Quality

Quality has everything to do with meeting the performance guarantees that the Owner has asked for and that the Contractor has given in the listing of performance guarantees under the contract. They have to do with a certain output resulting from the use of a defined feedstock under given specifications with a defined maximum energy consumption and e.g. a maximum allowable expulsion of NOx in the exhaust gasses of the plant.

Quality also has to do with the quality of the supplied hardware. Some Owners do not want hardware supplied from certain regions of the world because of quality concerns; some Owners wish to approve specific vendors or suppliers. To prevent approval issues with regard to vendors having a detrimental effect on the schedule (you cannot place orders until a vendor has been approved) the Contractor would do well to compile a list of vendors from whom he wants to procure goods and materials (or services) and ask the Owner to approve the vendors on the list. The Contractor removes the vendors that the Owner will not approve from the list. This results in an (approved) vendor list, attached to the contract, and placement of orders with these vendors will not need any further approval from the Owner.

Performance is measured with a pre-agreed method during a performance test in which the plant runs for 72 consecutive hours. Successful completion of the performance test means achievement of provisional acceptance, a common milestone in Oil and Gas contracts. One should be aware of the fact that under an Engineering and Procurement ("E+P") contract the moment of provisional acceptance may come much later than the Contractor's last delivery under the contract as the Contractor does not perform the construction and he does not have any influence on when the construction will actually take place.

The performance milestone is called provisional acceptance as the Contractor must also guarantee the installed hardware during a contractually agreed period, often two to three years. After expiry of the guarantee or warranty period, Final Acceptance is reached and the Contractor will have fulfilled all his obligations under the contract.

The contractual provisions with regard to warranty on the hardware should always contain a final cut-off date to prevent "evergreen" warranties, meaning warranties that are given on replacement parts that failed to perform. If the warranty on the replacement part is again the length of the warranty period then theoretically the warranty period would never be over. For further reference on this issue please see again Chapters 5 and 6.

The commissioning of the plant assures that all systems and components of an industrial plant are designed, installed, tested, operated and maintained according to the operational requirements of the Owner. However, as soon as the plant is operational and produces, many Owners refrain from executing the performance test(s), as such a test takes time away from production. The Owner will state that the plant is working, which means it is producing and, above all, making money.

However, the production conditions will most likely not comply with the agreed performance test conditions. Where the Owner may have a (short term) rationale not to stop production and to calibrate the plant to the agreed performance test settings and conditions, the Contractor cannot just accept the forfeiture of the performance test, as that would prevent him in a later stage to prove that the plant was designed, engineered and constructed in accordance with the contract. Thus, if within a certain period after commissioning, the performance test does not take place and this is not due to the Contractor, the plant should be deemed provisionally accepted, at which time the warranty period begins to run for the contractually agreed duration. The Contractor should have a provision in the contract that, in the event that during the Defects Liability Period faults in the Works arise, the Contractor must have the right to scrutinise the operational data of the plant about the period prior to the fault arising as the Owner could have used the plant beyond the specifications in order to increase his profit.

1.11 Damage and/or loss; injury or death

Accidents that happen on site that cause damage and or loss to property or, even worse, injury or death to persons can be a reason for lengthy litigation leaving particularly the victims or their next of kin in a very uncertain situation for many years as in this kind of litigation it may take years to unravel what or who exactly caused that accident and who is therefore liable for its consequences. As such risks are generally insured, such litigation will take up a lot of management time that should be used for other issues. This is why in the oil and gas industry the Owners and Contractors have invested heavily in developing, teaching and maintaining Health Safety and Environment (HSE) standards, to prevent accidents from happening. Where in heavy civil engineering accidents are still very common, they are much less so in the oil and gas industry, thanks to such HSE programs.

It seems that in heavy civil engineering for infrastructure Owners, in some countries, are not keen to make money available for the implementation of HSE programs. Hence the financing of such programs must come from the Contractor's budget (read: margins) and the margins in that industry are simply not high enough. Where in the oil and gas industry the Owners may vote with their feet by simply refusing to buy the Owner's products, such sensitivity of Owners (most probably: public authorities) in the heavy civil engineering, in the author's home country at least, does not seem to exist, which in the author's mind is very strange as a primary task of public authorities is to keep "the people" safe.

Accidents however may still happen in spite of everyone's good efforts to prevent them. To avoid ugly and lengthy litigation it is very common to have a *mutual hold harmless* and indemnity provision ("MHH") in the contract on a knock-for-knock basis.

Basically, it says:

> I will take care of my property and my people in exchange for you taking care of your property and people, irrespective of negligence or a breach of duty and we will get our insurers to waive subrogation on this matter.

As long as the waiver is reciprocal, insurers accept this.

It is important that the MHH is extended over each of the party's "group" so including: employees, agents, representatives of its suppliers, Contractors, consultants, subcontractors, co-owners and affiliated companies.

The knock-for-knock concept leaves responsibility for the loss or damage with the party that suffers it, i.e., the party that took or received the hit or "knock". Shipowners and their insurers adopted the concept during World War II in the case of night-time collisions between vessels in the same convoy where it was generally impossible to determine sole negligence by one of the parties, particularly while under attack from the enemy.

Within the oil and gas industry it is common for the engineering Contractors to want an MHH on a knock-for-knock basis. Often Owners have different ideas and generally wish to exclude gross negligence and wilful misconduct from the knock-for-knock. As practically no jurisdiction will allow exclusion of liability in the event of gross negligence or wilful misconduct, such a carve-out should generally be acceptable to the Contractor.

Knock-for-knock liability is less prevalent (but not necessarily less important) in downstream petro-chemical contracts than, for example, in offshore and subsea contracts (see Section 3.9).

There is a distinct difference between an assumption, exclusion or limitation of liability per se on the one side and the positive duty of a person or entity (the "indemnitor") to protect and hold harmless another person or party (the "indemnitee"), and a contract should preferably include both types of duty. A waiver of subrogation by the insurer of each party is necessary, which insurers are prepared to give if the obligation to give the waiver of subrogation is reciprocal. It saves years of litigation to have established who was at fault and caused the accident that caused the damage or the injury or worse the fatality.

A contractual provision for the mutual hold harmless could look like this:

> CONTRACTOR shall hold harmless and indemnify OWNER GROUP against all liabilities, claims, judgments, demands, damages, losses, costs and expenses arising directly or indirectly out of:
>
> 1 Injury to (including sickness) and death of CONTRACTOR GROUP's employees, agents, representatives, including injury to (including sickness) and death of the employees, agents, representatives of its suppliers, contractors, consultants, subcontractors, co-owners and AFFILIATES;
> 2 Loss of or damage to CONTRACTOR GROUP's own property and to the property of its own employees, agents or representatives, including loss or damage to the property of its suppliers, contractors, consultants, subcontractors, co-owners and AFFILIATES;
> 3 Pollution by CONTRACTOR GROUP originating and/or emanating from its property, equipment and spread; ***whether or not caused or contributed to by*** the negligence or other breach of duty of any of the OWNER GROUP. CONTRACTOR consequently waives, and shall cause their insurers to waive, all right of recourse against OWNER GROUP in respect thereof and shall save and hold OWNER GROUP harmless from and against any such liabilities, claims, judgements, demands, damages, losses, costs and expenses in connection therewith, arising thereof and/or relating thereto.
>
> OWNER shall hold harmless and indemnify CONTRACTOR GROUP against all liabilities, claims, judgments, demands, damages, losses, costs and expenses arising directly or indirectly out of:
>
> 1 Injury to (including sickness) and death of OWNER GROUP's own employees, agents, representatives, including injury to (including

sickness) and death of the employees, agents, representatives of its suppliers, contractors, consultants, subcontractors, co-owners and AFFILIATES;

2 Loss or damage to OWNER GROUP's property and to the property of its own employees, agents or representatives, including loss or damage to the property of its suppliers, contractors, consultants, subcontractors, co-owners and AFFILIATES;

3 Pollution originating and/or emanating from property, equipment and spread other than from CONTRACTOR GROUP's property, equipment and spread;

whether or not caused or contributed to by the negligence or other breach of duty of any of CONTRACTOR GROUP. OWNER consequently waives, and shall cause their insurers to waive, all right of recourse against CONTRACTOR GROUP in respect thereof and shall save and hold CONTRACTOR GROUP harmless from and against any such liabilities, claims, judgements, demands, damages, losses, costs and expenses in connection therewith, arising thereof and/or relating thereto.

The parties shall cause their respective insurers to waive the insurer's right to subrogation.

Insurers will waive their right to subrogation if the insurers of the other party will do the same; they are not interested in expensive and long litigation about who is liable for an accident. Such liability is mostly very hard to prove in a complex situation like a project.

The word "GROUP" needs to be defined in the Definitions provision of the contract as:

> "GROUP" shall mean either OWNER or CONTRACTOR and their respective co-venturers, co-owners, and its and their respective AFFILIATES, personnel, contractors and subcontractors and its and their respective directors, officers and employees.

The liabilities with respect to loss or damage to the Works and third parties are often, on the contrary, fault or negligence based. Most non-lawyers/non-insurers and even many lawyers are more familiar with fault/negligence-based liability than with knock-for-knock liability.

On the other side of the liability and indemnification spectrum is the concept of strict liability. It means that no matter how the damage was caused and who caused the damage one of the contractual parties will contractually have to accept liability for the damage even if caused by the other party or a third party. The following excerpt shows strict liability; the crucial wording is marked in italics.

> The CONTRACTOR shall assume all liability for and shall defend, indemnify and hold OWNER, any AFFILIATES OF OWNER and TECHNICAL ADVISER free and harmless from and against any and all claims, demands, actions or proceedings

> from third parties arising out of the execution of the WORK *howsoever arising and whether or not caused or contributed to by negligence or breach of duty* on the part of OWNER, any AFFILIATE OF OWNER or TECHNICAL ADVISER up to an amount of five million US dollars ($5,000,000) per incident or event and unlimited as to the number of incidents. In excess of this level, liability shall be determined by reference to applicable law.

Insurance is not a substitute for a contractual allocation of risk. In other words: first one should decide whether or not one will accept a risk and once the risk is accepted one may wish to mitigate the risk by insuring it. Insurance is an indemnification agreement. So, insurance can be a risk mitigation factor for the party accepting the risk; however, it should not be a factor in the decision to accept a contractual risk.

1.12 Dispute resolution

Once an issue under a contract develops into a full-blown dispute, it is important that it is resolved as soon as possible, yet in a just and fair manner. The problem with going through the judicial system in almost any country in the world is that it is very, very slow, not in the least because judges are lawyers and usually have no particular knowledge about engineering Contracting or construction in general.

The judicial system has to resort to the use of expert witnesses who are either court appointed or nominated by the litigating parties. That is a time-consuming process and its outcome is mostly unpredictable. Engineering Contractors are project-oriented organisations and are not looking forward to litigate on any issue for five to eight years.

The alternative is arbitration. The advantage of arbitration is that arbitrators are experts in their field themselves and thus far better equipped than legal scholars like judges to grasp the intricacies of a technical dispute. In international contracting it is important to be sure that the arbitrators will be un-biased and cannot be influenced by one of the parties. So, the preference is ICC arbitration or reputable arbitration institutes in certain countries. As a dispute going to arbitration will be a serious and sizable dispute it is imperative that three arbitrators rather than just one judge the dispute. It offers a better guarantee that an in-depth debate will take place on the merits of the case.

1.13 Takeaways

This chapter lists the main causes of disputes in construction projects and explains why they arise, how disputes can be prevented by doing the things that lead to disputes better and how to do them better.

Problems, possibly developing into disputes, often originate in relation to:

1 (Supply of) Information is of vital importance as a stable foundation for the project. Lack of, or faulty, information creates serious problems.

Owner should invest in a comprehensive investigation to gather complete and correct information that the Contractor may rely upon.

2 Scope of work and the functional specifications therein comprise what the Contractor must produce, and therefore it needs to be a comprehensive description.

3 (Lack of) Access to the site will prevent the Contractor from performing his obligations under the contract.

4 Clash of personalities in Project Management on the Contractor's and on the Owner's sides will cause a climate for disputes that will be detrimental to the project's objectives. The upper management of both Owner and Contractor must get together and decide which person is to be replaced.

5 Price, when it is a lump sum price, entails that the Contractor has all the risk with regard to correct pricing of all the work, materials, etc. for the project.

6 Payment intervals decide whether or not the Contractor has the risk pre-financing the project, and the best thing is to plan payment on achieved milestones so that the Contractor does not have to bear that risk and can have a neutral or slightly positive cash flow. Payment security for the Contractor comes from a confirmed letter of credit, confirmed by the Contractor's own bank.

7 Risks originating outside of the parties' control, e.g. unforeseeable physical conditions, *Force Majeure*, etc., may cause delay and additional cost. The NEC 3 approach of defining physical conditions which have such small chance of occurring that it would be unreasonable to have allowed for them in the contract price as a contingency is a better way of dealing with unforeseeable physical conditions than apportioning risk to the Contractor for what an experienced Contractor could/should have foreseen and that such foreseeability by an experienced Contractor would deny him any claim for extra money and extra time as the presence of such conditions increases his scope of work.

8 Delay: the Contractor must make sure that where he is dependent on the Owner for his progress, e.g. because of Owner approval of design work or delivery of Owner-supplied hardware or work, such performance by the Owner is subject to time limits. With regard to approval of design work, it is common to set a time limit of ten to 20 days (working days or calendar days) and the contract must then provide that if the Owner does not approve within that time limit, the design work shall then be deemed approved.

9 Quality (Performance) has everything to do with meeting the performance guarantees that the Owner has asked for and that the Contractor has given in the listing of performance guarantees under the contract. They have to do with a certain output resulting from the use of a defined feedstock under given specifications with a defined maximum energy

consumption and e.g. a maximum allowable expulsion of NOx in the exhaust gasses of the plant. Quality also has to do with the quality of the supplied hardware.

10 Damage and/or loss, and injury or death may cause very long and costly litigation and therefore it is best to agree on a mutual hold harmless provision on a knock-for-knock basis with a waiver of subrogation by the insurers of the parties, which the insurers will happily give if the waiver is reciprocated.

11 Flexibility in dispute resolution: to be able to achieve the project's objectives it is of paramount importance to resolve such disputes quickly through the use of Dispute Adjudication Boards.

The following chapters will take an in-depth look at the sequential phases of a project and the main causes of disputes.

CHAPTER 2

The conception of an idea for a project

2.1 Introduction

This chapter is all about how a project is to be treated on the Owner's side from the moment the idea for the project comes up, and how to be realistic about the project's revenue after completion and the cost and necessary time for the design and construction.

2.2 The beginning

Every project starts with someone having an idea for something to be built that would enhance and improve the Owner's business or, in the case of a public authority, would enhance and improve the authority's service to the public. An idea could also manifest itself in a Contractor's organisation because something new has been conceived that can be beneficial to the Contractor's Clients. This happens frequently in the field of the generation of "green" energy, but it can also happen in infrastructure where the public is e.g. frustrated by daily traffic jams or other disturbances causing extensive delay in commuting and someone has an idea how to resolve the "bottle neck" and increase the flow of traffic.

Once the idea is there, the Owner needs to establish if the idea is feasible, both technically and financially.

Good management is all about enabling people to excel at what they are good at, without the noise of personal interests, career opportunities, bonuses, etc. It is just enabling people to learn and grow and to be effective in what they do, in a harmonious environment where they know they will be appreciated for a job well done. In relation to projects: to get people to focus on the success of the project above anything else, paraphrasing John F. Kennedy: "Ask not what the project can do for you, ask what you can do for the project!"[1] That is what management must make clear to their staff; achieving the project's objectives as defined and communicated to them is the only thing that matters and that this is what staff will be evaluated on. It is of paramount importance to avoid individual

1 John F. Kennedy's inaugural address 20 January 1961: "Ask Not What Your Country Can Do for You, Ask What You Can Do for Your Country".

"kingdoms" arising within the project staff and the home office staff to ensure that all staff are focusing on delivering a successful project.

2.3 Feasibility study

A feasibility study must be launched and executed with due diligence, covering both the technical side and the financial side. Once the study is completed and it provides sufficient detail in all aspects that indicate that the idea is indeed truly feasible, the organisation stands before a go/no go decision. This will not be the final go/no go decision; there will be more in subsequent phases and the decision should limit itself to: Can it be done? Will it be beneficial, and is it within the organisation's strategy? The answers to these questions can be found by first executing a basic level "due diligence" investigation. By involving technical and financial disciplines one can at least establish if the original idea has realistic potential. In this phase it is important to resist the feeling of enthusiasm for one's own idea and first establish a basic level of feasibility of the idea before making the decision to continue. If not, one runs a potentially serious risk of wasting significant amounts of money on an idea that, later on, proves either technically and/or financially not feasible.

This is true for both the private realm and for the public realm.

Research by Flyvbjerg, Cantarelli, Molin and van Wee[2] and several others has shown that 70% of cost overruns originate in the pre-construction phase and the longer that phase takes the higher the cost overruns will be.

The desire to make political gain from a high-profile infrastructure project and pushing it through while rationalising that during the various phases of the project things will be taken under control and kept under control, can result in serious cost overruns. That philosophy is one of the roots of so many public projects ending up in serious cost overruns and delays simply because the preparation of the project falls significantly short of the level of realistic planning and costing necessary for a successful completion of a project. The necessary data will not be gathered because:

- There is not enough time, and
- There is a reluctance to spend money on a due diligence investigation to gather the facts that are necessary for a smooth transition from conceiving the idea to the procurement of the project and the selection of the Contractor who is to design the project, procure the materials necessary and construct the designed works, test them and hand over the completed works to the Owner.

Once an objective investigation has been executed showing that the project is indeed deemed feasible, both technically and financially, at least in the sense that funds can be made available and that should the price submitted by the

2 C. C. Cantarelli, E. J. E. Molin, B. van Wee & B. Flyvbjerg: "Characteristics of Cost Overruns for Dutch Transport Infrastructure Projects and the Importance of the Decision to Build and Project Phases". *Transport Policy*, 22, 2012, pp. 49–56.

tendering Contractors be higher than expected there is a strategy to acquire additional funding, then the decision can be made to go ahead and formulate the Owner's Requirements, their accompanying documents containing the necessary information, etc., describing the exact functional specifications of the facility to be built, and the invitation to tender can be launched.

Public project procurement falls outside of the scope of this book, as public procurement rules differ significantly from jurisdiction to jurisdiction. In the private realm the message is: "Treat every tenderer the same as the others and above all: treat them fairly", as also is the credo in public project procurement but in the public realm this credo is generally consolidated in strict public procurement rules.

Optimal preparation is a key to success and though there are many more keys to success good preparation is the primary key to success. The Dutch ABN-AMRO Bank has published a report in the Dutch language, expected to be translated into English, on the causes of cost of failure in the construction sector named *Verspilde Moeite* [in English: "Wasted Effort"]. It is a report on several interviews with Engineers, Contractors and Management Consultants by ABN-AMRO construction sector specialists. The conclusions and recommendations of this excellent report[3] are 1. a good preparation is "half the work"; 2. cooperation will get you better results; 3. experience makes all the difference; 4. innovation and standardisation are paramount; 5. share knowledge and learn from mistakes.

1 **Good preparation**: more time must be spent on the start-up phase when the design is made and the project is being prepared. Time must be devoted to: the Owner's demands (requirements); realistic planning; and identification of risks at the start of the project which will result in less uncertainty and cost in the execution phase. A good preparation will prevent variations ordered by the Owner in the execution phase which will cost time and money.

2 **Cooperation**: good cooperation is essential and must be optimised resulting in knowledge input and sharing, better involvement and less misunderstanding between the parties. The interviewees named cooperation with regular partners as an important step to reduce the cost of failure. Alliancing between the primary parties in a project supports the good outcome of the construction process and the planning.

3 **Experience**: training of influx of new young employees by experienced members of the staff who have a "knack" for transferring knowledge to the less experienced will pay for itself by the reduction in the cost of failure. So, spending time end effort in training of the less experienced is definitely worth it.

4 **Innovation and standardisation:** process optimisation and process innovation like more efficient planning, digitising, and automation help in the prevention of failures.

5 **Knowledge sharing and learning from mistakes:** learning from mistakes requires open communication about those mistakes with one's partners. Focusing on innovation now will reduce failures in the future as will

3 "Verspilde Moeite" by ABN-AMRO Bank; Petran van Heel, Madeline Buijs & Casper Wolf.

being less dependent on the knowledge and experience of a few experienced staff members.

Obviously, the title of the report "Wasted Effort" relates to all the efforts to initiate a project *without* implementation of these five recommendations by the interviewees in the report.

One could add that all or most of these recommendations would apply to Owners as well.

All too often a comprehensive due diligence investigation into the financial and technical feasibility is avoided as not to spend substantial amounts "to find out that the project is not possible" or to "waste" budget for the project on studies. However, such a study is absolutely needed to compile the Owner's Requirements comprising all data and information Contractors need to prepare their tenders for the project.

2.4 Decision-making after the due diligence

Beyond this initial phase, completion of the due diligence investigation, the work of the Owner is certainly not yet done. There is a disturbing tendency to shift all the risks towards the Contractor by employing the word "Turnkey" in a construction contract and then expect that this will shift all the risks towards the Contractor. This should be classified as "Total Nonsense", the word "Turnkey" is an indication of the extent of the scope of work to be executed by the Contractor and not a "cheap" way of ridding oneself as an Owner of all risks and thus cost overruns. Shifting all the risks to the Contractor is the equivalent of someone getting into their car in the morning and saying: "If I have a collision today, it will be the other person's fault and liability".

The FIDIC Silver Book, one of the primary examples of standard conditions for Turnkey contracts, contains a caution to only use the Silver Book if all possible risks related to the project are known, as all these risks are allocated to the Contractor and this caution can be found in the introductory note to the first edition published in 1999 where it states:

> These Conditions of Contract for EPC/Turnkey Projects are not suitable for use in the following circumstances:
>
> - If there is insufficient time or information for tenderers to scrutinise and check the Owner's Requirements or for them to carry out their designs, risk assessment studies and estimating (taking particular account of Sub-Clauses 4.12 and 5.1).
> - If construction will involve substantial work underground or work in other areas which tenderers cannot inspect.
> - If the Owner intends to supervise closely or control the Contractor's work, or to review most of the construction drawings.

- If the amount of each interim payment is to be determined by an official or other intermediary.

Obviously the first bullet point is very significant though it states "if there is insufficient time or information to scrutinize".

Apparently, many Owners do not read the introductory note and use these conditions of contract in spite of a lack of data and information.

The data resulting from the due diligence investigation form the virtual foundation of the project upon which the Contractor will have to build the project and the data will have to be incorporated in the Employer's Requirements. The Contractor will have to scrutinise the Employer's Requirements, and if there are any unclear issues he will have to ask for clarification of such issues. The Contractor will need ample time for scrutinising, and it is of great importance to the well-being of the project that all the tenderers get sufficient time for scrutinising. Demanding that the tendering Contractors have identified all errors in the Employer's Requirements at or before submittal of their tenders and that if they fail to do so they will forfeit their right to claim extra cost and time for completion is a recipe for creating multiple and severe disputes.

Professor Bent Flyvbjerg, Oxford University, has done extensive research into why so very many projects end up with cost overruns and delays and presents his paper: "From Nobel Prize to Project Management: Getting Risks Right".[4] He sees on the basis of his research as a major source of risk in construction:

1 Inaccurate forecasts of project costs, project demand and other impacts, which lead to inaccuracies in Project Management;
2 Bias and Strategic Misrepresentation.

Why the inaccuracies? People have been building projects for many years, so why are there still inaccuracies in forecasts of project cost? Use of unreliable data or outdated data, in other words: the technical explanation?

Bias has a psychological component: optimism – people believe things will turn out fine even though the data tell them differently. This component one sees when political and organisational pressures are low.

It also has a political component: strategic misrepresentation – overestimating benefits and underestimating costs. This component one sees when political and organisational pressures are high.

Flyvbjerg sees as a solution to the cognitive bias, to which he refers as the Planning Fallacy (underestimating costs, completion time and risks of planned actions, and overestimating benefits of those actions), a method named: *Reference Class Forecasting*, which requires three steps:

1 Identification of a relevant class of past similar projects, which class must be:

4 Bent Flyvbjerg: "From Nobel Prize to Project Management: Getting Risks Right". *Project Management Journal*, 37, 3, August 2006, pp. 5–15.

   a broad enough to be statistically meaningful and
   b narrow enough to be truly comparable;

2. Establishing a probability distribution for the selected reference class: credible, empirical data to make statistically meaningful conclusions;
3. Comparing a specific project with reference class distribution to establish the most likely outcome for the specific project.

This will result in a statistical distribution of outcomes from the class of reference projects, so a "best guess" of cost and completion time based on the average of the reference class projects. All this should be part of the Owner's due diligence investigation with regard to the project he wants.

2.5 The procurement of the project

Once the Owner has finished the due diligence investigation, he will write or have a consultant write the Employer's Requirements which are the document(s) setting out the Employer's Requirements for the Works, as included in the contract, and set out in an Appendix (if any), and any variations thereto. The Employer's Requirements include, without limitation, documents specifying the scope, purpose, functions, time frame for construction, and/or design and/or other technical criteria, and/or conceptual elements for the Works.

If the project is quite substantial it makes sense for the Owner to start a pre-qualification procedure to select Contractors on the basis of experience with projects of similar size and complexity that they have completed to the satisfaction of the Owners of those projects expressed in a letter from said Owners. The Contractors that pass the pre-qualification criteria will then be invited to tender for the project.

The Employer's Requirements are to be used by the Owner in the invitation to tender to the Contractors together with the proposed form of contract. They must describe what the Owner wants, what he intends to use it for and how he intends to use it (intended purpose), all to give the Contractor a clear picture of what he must design and build while considering if the Owner will use the Works eight hours per day, or 12 hours per day or even the full 24 hours per day. This will have consequences for the design; e.g. if the Works are a power plant that must be online 95% of the time and the Owner changes its mind to 98% of the time online, substantial parts of the Works may have to be doubled as maintenance of the plant takes more than 2% of the time. This will be discussed in the next chapter.

2.6 Takeaways

In budgeting the project, the Owner and his staff must be very thorough and careful. Time and again projects are budgeted incorrectly causing major cost and time overruns.

The major sources of risk are:

> Inaccurate forecasts of project costs, project demand and other impacts. Inaccuracies in management of the project in early stages because of bias, being optimism leading to self-deception and strategic misrepresentation or, in plain English, lying about the budget and time needed for completion of the project just to get the decisions to go ahead with the project made.

CHAPTER 3

The Employer's Requirements, the conceptual design and the contract conditions

3.1 Introduction to the Employer's Requirements

The Employer's Requirements form the very foundation of the entire project and are essential, as foundations are for the stability of a building. They are indispensable and must be put together with great care, by people with knowledge and understanding of the various contents. The big challenge for management, both within the Owner and the Contractor, is to know their own people in-depth and to know how each individual approaches and assesses the issues connected to the Employer's Requirements. Do they approach and evaluate with an open mind or do they act from a mind that is already made-up, are any of them biased either negatively of positively with regard to the success of the project or is their mind-set neutral and are they allowing the facts and figures determine their opinion on the project? This is an essential issue both within the Owner's organisation as well as within the tendering Contractors' organisations.

Particularly for industrial/technical installations the Employer's Requirements can comprise 100+ pages of information. As each project has its own individual issues, the Employer's Requirements of an earlier similar project cannot just be employed for a subsequent project.

An example of the Employer's Requirements broad contents list for a chemical plant:

1 Definitions,
2 Conceptual design,
3 Scope of project,
4 General requirements,
5 Particular requirements,
6 Facility requirements.

The Definitions give the defined terms/words that are used and their meaning as to avoid misunderstanding of their meaning in the document.

3.2 The conceptual design

The conceptual design offers a broad description of the project, the facility to be built, its location, its general purpose and specific purposes, location of the site, site conditions, feedstock particulars, site ambient conditions, sketches (if any) etc.

Scope of the project offers a description of the scope of works, all work (foundations, structures and buildings), plant, labour, materials, construction plant, Temporary Works, auxiliaries and accessories (spare parts, etc.) and performing all work and required for design, fabrication, testing, transport and delivery to site, installation and commissioning, finishing, functional, performance and acceptance testing, final inspection and Punch List resolution, making good defects during the Defects Liability Period.

3.3 General requirements

General requirements describe the standards the Works must adhere to, the safety and management legislation, engineering drawings and documents, status indication of drawings and requisitions, as-built drawings, project control, planning, facility design life, design stage, construction stage, quality assurance, inspection and test records, subcontractor information, environmental constraints, site security, inspection and testing, commissioning tests, training of operational staff, etc.

3.4 Particular requirements

Particular requirements describe requirements with regard to output, noise abatement, vehicular access points, vents, stack(s), hours of use per day, etc.

3.5 Facility requirements

Facility requirements describe the feedstock composition and supply, reception and storage facilities, facility layout, feedstock transportation data, feedstock weighing, etc., offloading bays, sampling plant, emissions, etc.

All in all, a plethora of information and the better all this is put together, the better the chances for a good project as all this information makes it clear to the Contractor what is expected of him. Where the Contractor finds gaps in this information he should warn the Owner. Often Contractors don't warn, particularly not during the tendering stage as such gaps might offer him an opportunity to make good on losses he incurred because of his own oversight of certain issues. If the applicable law does not put a legal obligation on the Contractor if he discovers a certain lack of information as civil law commonly does, it is recommendable, once the project is awarded, to put a contractual obligation on the Contractor to do so. The same goes vice versa for the Owner, who should warn the Contractor of any gap in information or mistake he discovers.

3.6 The tender phase

From the Employer's Requirements, the tendering Contractors must be able to make a complete mental image of what the end project must encompass and what it must be able to perform. This is essential for the tendering process as it is the only way the Owner can compare the various tenders received from the tenderers and the only way he will know that he can compare the individual tenders and which one answers the best to the award criteria and which tenderer he must select as the Contractor for his project.

It is not uncommon to have a Front-End Engineering and Design ("FEED") made first and to have Contractors tender for the opportunity to do the FEED before putting the entire project on the market. Then the Owner will have a well-documented and detailed plan for the project including the rough identification of the main costs of the project. As usually there is close communication between project Owners, Project Operators and the Engineering Contractor, the project requirements can be worked out well, which may prevent significant changes during the execution phase of the project and thus additional cost being incurred and additional time being required for the completion of the project. The FEED, if executed well by a qualified Contractor, will provide a good solid basis for the engineering, procurement and construction phase of the project and will provide the Owner a reasonable idea of the cost of the project so that he can organise a solid budget for the project. The FEED is in fact an elaborate feasibility study.

The FEED commonly has different stages referred to as "Front End Loading one, two and three" and as the discussions between Owner, Project Operators and the engineering Contractor progress they result in the following deliverables: Material Balance, Energy Balance and Project Definition (FEL I); Preliminary Equipment Design, Preliminary Lay Out, Preliminary Schedule and Preliminary Estimate (FEL II); purchase-ready major equipment specifications, definitive estimate, project execution plan, Preliminary 3-D model, electrical equipment list and line list. The project becomes ever more clear and this approach prevents overzealous optimism on the side of the Owner about the necessary budget and required time for the completion of the project. FEL I usually has a +/−30% accuracy; FEL II +/−20% accuracy; FEL III +/−10% accuracy with regard to the total cost of the project. It provides an excellent reality check and should reduce the occurrence of claims and disputes considerably as it also provides good information for the Contractors to tender on.

3.7 The contract conditions

The Owner has the option to either have a bespoke contract be put together by his in-house lawyers in combination with external lawyers, or use a standard form of contract, contract conditions drawn up usually by a professional organisation such as NEC, JCT, FIDIC, ICC, CRINE LOGIC (Cost Reduction In New Era) by LOGIC, a wholly owned subsidiary of Oil & Gas UK, etc.

Usually contract conditions consist of two parts:

1 The General Conditions, which define each contracting party's obligations and rights, and the rules by which they are to meet their obligations and perform the work they agreed upon.

2 The Special Conditions, which make the contract conditions *project specific*, e.g. this is where the following items are identified: law governing the contract, the ruling language of the contract, the Employer's Requirements, percentages of overhead cost and profit margin for the Contractor on extra cost incurred in certain events, when a party will be relieved from its obligations, or not, etc.

The best contract conditions are balanced conditions, because they do not favour one party over another but allocate risk to that party that is best able to either control a risk materialising or best able to price its effects when it materialises.

Unfortunately, financiers involved in projects to be executed under a project finance scheme mostly do not see it that way. They prefer to keep the project's risks away from the Owner, who borrows the money from lenders after having proven that the project once finished will make a return on investment that is acceptable to the lenders, represented by a facilitating financial organisation, usually a bank that has produced a prospectus that states the cost of the project; the time necessary for realisation of the project; the revenue the finished project will yield; and the contract conditions to be used. The proof obviously is on paper only; the financial model shows that the cost of realising the project and the desired return on investment will be compensated by the project's earnings once the project is operational. Thus, in project finance there is a drive to allocate most risks towards the Contractor and that is generally not good for achieving the project's objectives and thus the outcome of the project as well as the return on investment.

3.8 Owner's obligations

There are several elements to the Owner's obligation to allow the Contractor to carry out and complete the Works. These include:

- Access to the site,
- Necessary permits or approvals,
- Specified start date,
- Provision of instructions and drawings,
- Provision of information,
- Duty of non-interference,
- Payment of work done,
- Acceptance of the Works.

3.8.1 Access to and possession of the site

In the first place, the Owner has to make sure that the Contractor has access to the site. The contract may actually give the Contractor possession of the site. Failure of the Owner to give access to or possession of the site will expose him to claims from the Contractor for extension of time for completion and/or for damages for delay or increased costs. One of the implied obligations of the Owner in providing access to the site is that the site is suitable for the Contractor to bring his equipment and resources that are required for the work and subsequently have them on the site. In addition, there is an implied obligation of the Owner to ensure that the Contractor is made aware of any restrictions that may occur, and of any underground obstructions.

The invisible nature of the sub-ground conditions, particularly on projects that require substantial foundations for heavy equipment, can frequently cause claims being raised by the Contractor at an early stage.

However, the Contractor is responsible for his own interpretation of the data given to him by the Owner.

The duty to give access and availability of the site to a Contractor who is involved at a later stage of the project also implies that the Owner has to ensure that any previous Contractors – such as a piling or groundworks Contractor – whose work on site precedes the later Contractor is finished on time so as to allow the later Contractor to start his work.

3.8.2 Permits and approvals

The Owner has to provide the necessary permits and approvals, or at least assist the Contractor in obtaining these on his behalf. It should never be the contractual obligation of the Contractor to obtain these necessary permits and approvals himself, and such an obligation should always be omitted from a contract.

On the other hand, it is valid for the contract to state that the Contractor is not entitled to claim if his performance is delayed because the necessary permits or approvals are not obtained in time by the Owner.

3.8.3 Specified start date

This may at first glance be glaringly obvious, but be aware of the difference between a date for the start of the construction of the Works and the date of a contract coming into force. These are two distinct things. A contract may often declare a specific date for the work to commence on site, but this is different from when a contract becomes effective.

Similarly, be aware of completion dates. Are they calculated as X number of months from the contract coming into force, or is there a specific date by which completion must be reached? Agreement on the start date of the contract is fundamentally important, for, as we will see later, this is the start point of how we can measure any delay that may occur on the project. If you don't know the starting point to begin with, it will be impossible to properly measure the extent of delay.

Sometimes the start of a contract coming into force is determined by the completion of a number of commercial tasks by the Owner, such as the signing of the contract, the receipt by the Contractor of an advance payment, final approval of Hermes export credit insurance, final approval of the relevant authorities in the country where the project is to be built and payment guarantees issued by the Owner. Only when the last of the stated preconditions falls into place, can the contract come into force and be effective. If the Owner is responsible for some unreasonable delay in the implementation of any of these preconditions, then the Contractor is normally entitled to present a claim for additional costs arising out of such delay.

3.8.4 Instructions and drawings

The Owner is often required to provide the necessary instructions and drawings to allow the work to proceed as planned. Sometimes, where the Turnkey Contractor's engineering drawings are subject to the Owner's approval (but not where they are submitted for his review!), and where the Owner takes longer to approve the drawings than set out in the contract, the Contractor may be entitled to claim for an extension of time and compensation for any extra costs incurred.

3.8.5 Information

It is important that the Owner provides any necessary information to the Contractor as required and in a timely manner. This is of particular importance in regard to any extra work that needs to be carried out (or to any contracted work that is to be deleted from the project). Any late information from the Owner to the Contractor may give rise to a claim for delay and disruption due to out-of-sequence working where the changed requirements of the Owner are made known to the Contractor far later than they should have been.

In addition, some contracts require the Owner to supervise the Works as they are being constructed, and if the Contractor is told about corrections that need to be made to his work later than they should have been, then again the conditions may be right for a claim for further delay and disruption.

3.8.6 Duty of non-interference

The obligation of an Owner not to interfere in the Contractor's prosecution of the Works is a classic example of an implied contractual term. In other words, the contract may not specifically declare that the Owner must not interfere with the Contractor's progress, but it is implied within the contract that the Owner may not interfere.

The Owner may have the authority within the contract to decide a certain method of construction or the use of certain materials, but this right cannot be extended to obstruct or prevent the Contractor from executing and completing the Works in accordance with the Contract Schedule.

The Contractor has the right to carry out the construction work as he thinks fit, provided of course that safety is ensured and that no laws are broken.

Interference by the Owner can occur in the engineering stage of the contract, particularly in regard to specifications. Specifications can be characterised as either "design" (technical specifications) or "performance" (performance specifications).

Technical specifications describe the materials to be used in detail and the way in which the work is to be carried out. A design specification will give no discretion to the Contractor, who is required to follow it. Performance specifications, however, set out an objective or standard to be achieved, and the Contractor is expected to exercise his ingenuity and expertise in achieving that objective or standard of performance, using the methods that he considers to be appropriate and taking responsibility for that selection.

In the United States, a doctrine has developed stating that design specifications contain an implied warranty that if these are followed, an acceptable result will be produced. This implies that if the Contractor complies with the specifications then he will be relieved from all responsibility if the project fails to perform. However, the Contractor would still share some of the blame if it appeared that there was a design defect in the Owner's specification and if the Contractor, having noticed the defect, failed to draw the Owner's attention to it.

In Europe, there is a tendency to express specifications in terms of performance as functional specifications, setting out the goals to be achieved rather than the methods to be used. In this way, the Owner successfully shifts some of the risk on to the Contractors who are contracting more and more under "fitness for purpose" obligations.

In determining the responsibility of the contract parties, the extent of the Owner's expertise and knowledge is taken into account. Many public authorities and international companies have specialised services in-house, which allow it to produce detailed designs and provide the Contractor with sufficient detailed information.

In a Turnkey project, however, the Contractor is usually required to supply a project that will meet the contract's performance specifications and the performance guarantees. The Contractor can only be excused from his liability for defects if he warns the Owner in advance about the risks resulting from the instructions, and the Owner nevertheless insists on the project being built in accordance with his instructions.

3.8.7 Payment of (sections of) work completed

The contract conditions the Owner proposes to the Contractor must allow for payment of work done in relation to parts of the work done measured in milestones. The risk of the Contractor is of immense proportions if he is to receive payment when all work is completed, plus the fact that this would mean the Contractor has to pre-finance the entire project. Allowing the Contractor to divide the work in milestones and invoice accordingly will allow him to plan the milestones so that he maintains a neutral or slightly positive cash flow on the project giving him the comfort of knowing that he will not have the risk of non-payment or of late payments.

3.8.8 Acceptance of the Works

Time is of fundamental importance in construction contracts. Projects are normally to be completed by a certain date, and in the case of commercial projects, this usually means as soon as possible. If there is no fixed date for completion, but perhaps a defined contract period in a set number of days, weeks or months, then the parties must agree as to the final completion date.

In the event of delay that prevents the completion date from being reached, the Contractor is normally liable for damages (by what we call liquidated damages for delay) unless he can prove that the delay has been caused by the Owner himself. A case of being guilty until proven innocent!

In some countries – such as Belgium, for example – liquidated damages can only be enforced to the extent that they are a genuine pre-estimate of the extent of the damage that the Owner suffers. If the amount is too high, the judge will refuse to apply the liquidated damages clause. Also in The Netherlands the judge or arbitrator can mitigate the liquidated damages to an amount equal only to the real damages suffered by the Owner.

The completion of the Works triggers the acceptance of the Works by the Owner. On many projects, the work is accepted in two stages:

- At provisional acceptance, where the Works are inspected and defects are recorded, while giving the Contractor the opportunity during a certain period to correct these defects; and
- At Final Acceptance, which is at the end of the Defects Liability Period, where the Owner gives his final approval and where the Contractor is discharged of his responsibilities under the contract.

The acceptance of the Works indicates that the Owner is satisfied with the Works, therefore when the Works have not been executed in accordance with the contract the Owner can refuse to accept them. The Owner is not allowed to refuse acceptance of the Works if only minor defects exist, and if the Owner unjustly refuses to accept the Works the Contractor can sue him and be granted acceptance. In Germany, the Standard Conditions of Contract (VOB/Bs. 12 no.4) explicitly provide for formal acceptance to be granted in the absence of the Owner if he fails to appear at the time due for acceptance.

In some contracts, often in the Middle East, Final Acceptance can be granted after a specific period of time, even though the project is not actually physically complete, provided that the Contractor can prove that a (usually lengthy) delay to the project is not attributable to himself. However, the issue of a Final Acceptance Certificate in these circumstances does not necessarily mean that the Contractor is discharged from his responsibilities nor that he can simply walk away from the project.

In Belgium, The Netherlands and France, the acceptance is the legal starting point for a special decennial liability, which means that for the next ten years the Owner can hold the Contractor responsible for defects that make a building unfit for its intended purpose. This decennial liability can also be found in particular contracts across the world, though not necessarily enforceable by law. In Germany the prescribed period for defects is generally limited to five years, while in the UK the issue of a "Certificate of Practical Completion" triggers the start of a "Defects Liability Period" of usually 12 months. Thereafter final approval of the Works is given in a "Certificate of making good Defects" and the Contractor remains liable for any hidden defects for various periods, dependent upon the law, for up to ten years.

Final Acceptance also triggers the Owner's obligation to pay the balance of the contract price and return any performance guarantees that remain outstanding.

As the initiator of the project, the Owner has a crucial role in the construction process. His opinion on his own role will determine many factors in the execution of the project. Moreover, the Owner can play a very important role in avoiding disputes by structuring the operation in such a way that the issues that are commonly known to give rise to problems or claims can be limited or excluded in the contract documentation. His concern, after all, is not only to achieve completion of his project on time and within budget, but also – if possible – to do so without claims and dispute resolution.

3.8.9 Liability for injury or death and for damages or loss of property

Accidents may happen and they do during the construction phase as work is being executed on the site. It is often extremely hard to assess who did what wrong and therefore is liable for the consequences of his wrongdoing. To avoid having to spend years in litigation and very substantial amounts in legal fees, it is a sensible thing for both contractual parties to provide each other with a mutual hold harmless indemnification on a knock-for-knock basis.

The knock-for-knock concept leaves responsibility for the loss or damage with the party that suffers it, i.e., the party that took or received the hit or "knock". Shipowners and their insurers adopted the concept during World War II in the case of night-time collisions between vessels in the same convoy where it was generally impossible to determine sole negligence by one of the parties, particularly while under attack from the enemy.

Within the oil and gas industry it is common for the engineering Contractors to want an MHH on a knock-for-knock basis. Often Owners have different ideas and generally wish to exclude gross negligence and wilful misconduct from the knock-for-knock. As practically no jurisdiction will allow exclusion of liability in the event of gross negligence or wilful misconduct, such a carve-out should generally be acceptable to the Contractor.

Knock-for-knock liability is less prevalent (but not necessarily less important) in downstream petro-chemical contracts than, for example, in offshore and subsea contracts (see Section 3.9).

There is a distinct difference between an assumption, exclusion or limitation of liability per se on the one side and the positive duty of a person or entity (the "indemnitor") to protect and hold harmless another person or party (the "indemnitee") and a contract should preferably include both types of duty. A waiver of subrogation by the insurer of each party is necessary, which insurers are prepared to give if thc obligation to give the waiver of subrogation is reciprocal. It saves years of litigation to have established who was at fault and caused the accident that caused the damage or the injury or worse the fatality.

A contractual provision for the mutual hold harmless could look like this:

> CONTRACTOR shall hold harmless and indemnify OWNER GROUP against all liabilities, claims, judgments, demands, damages, losses, costs and expenses arising directly or indirectly out of:
>
> 1 Injury to (including sickness) and death of CONTRACTOR GROUP's employees, agents, representatives, including injury to (including sickness) and death of the employees, agents, representatives of its suppliers, contractors, consultants, subcontractors, co-owners and AFFILIATES;
> 2 Loss of or damage to CONTRACTOR GROUP's own property and to the property of its own employees, agents or representatives, including loss or damage to the property of its suppliers, contractors, consultants, subcontractors, co-owners and AFFILIATES;
> 3 Pollution by CONTRACTOR GROUP originating and/or emanating from its property, equipment and spread; ***whether or not caused or contributed to by*** the negligence or other breach of duty of any of the OWNER GROUP. CONTRACTOR consequently waives, and shall cause their insurers to waive, all right of recourse against OWNER GROUP in respect thereof and shall save and hold OWNER GROUP harmless from and against any such liabilities, claims, judgements, demands, damages, losses, costs and expenses in connection therewith, arising thereof and/or relating thereto.
>
> OWNER shall hold harmless and indemnify CONTRACTOR GROUP against all liabilities, claims, judgments, demands, damages, losses, costs and expenses arising directly or indirectly out of:
>
> 1 Injury to (including sickness) and death of OWNER GROUP's own employees, agents, representatives, including injury to (including sickness) and death of the employees, agents, representatives of its suppliers, contractors, consultants, subcontractors, co-owners and AFFILIATES;
> 2 Loss or damage to OWNER GROUP's property and to the property of its own employees, agents or representatives, including loss or damage to the property of its suppliers, contractors, consultants, subcontractors, co-owners and AFFILIATES;

3 Pollution originating and/or emanating from property, equipment and spread other than from CONTRACTOR GROUP's property, equipment and spread;
whether or not caused or contributed to by the negligence or other breach of duty of any of CONTRACTOR GROUP. OWNER consequently waives, and shall cause their insurers to waive, all right of recourse against CONTRACTOR GROUP in respect thereof and shall save and hold CONTRACTOR GROUP harmless from and against any such liabilities, claims, judgements, demands, damages, losses, costs and expenses in connection therewith, arising thereof and/or relating thereto.

The parties shall cause their respective insurers to waive the insurer's right to subrogation.

Insurers will waive their right to subrogation if the insurers of the other party will do the same; they are also not interested in expensive and long litigation about which insured party is liable for the consequences of an accident. Such liability is mostly very hard to prove in a complex situation like a project.

The word "GROUP" needs to be defined in the Definitions provision of the contract as:

> "GROUP" shall mean either OWNER or CONTRACTOR and their respective co-venturers, co-owners, and its and their respective AFFILIATES, personnel, contractors and subcontractors and its and their respective directors, officers and employees.

3.9 Takeaways

1 The Employer's Requirements form the very foundation of the entire project and are essential, as foundations are for the stability of a building. They are indispensable and must be put together with great care, by people with knowledge and understanding of the various contents.

- The defined terms must be given a specific unambiguous meaning.
- The conceptual design gives a broad description of the project: the general purpose, the specific purposes, the location of the site, the site conditions, site ambient conditions, sketches (if any).
- The General Conditions describe the standards the Works must adhere to, the safety and management legislation, engineering drawings and documents, status indication of drawings and requisitions, as-built drawings, project control, planning, facility design life, design stage, construction stage, quality assurance, inspection and test records, subcontractor information, environmental constraints, site security, inspection and testing, commissioning tests, training of operational staff, etc.

- The particular requirements describe requirements with regard to output, noise abatement, vehicular access points, vents, stack(s), hours of use per day, etc.
- The facility requirements describe the feedstock composition and supply, reception and storage facilities, facility layout, feedstock transportation data, feedstock weighing, etc., offloading bays, sampling plant, emissions, etc.

2 From the Employer's Requirements, the tendering Contractors must be able to make a complete mental image of what the end project must encompass and what it must be able to perform. This is essential for the tendering process as it is the only way the Owner can compare the various tenders received from the tenderers and the only way he will know that he can compare the individual tenders and which one answers the best to the award criteria and which tenderer he must select as the Contractor for his project.

3 A good way of getting solid data for the Contractors to bid on is to have a Front-End Engineering and Design package (FEED) done, which can be seen as an elaborate feasibility study in stages named Fel I, Fel II and Fel III in ever increasing accuracy of the budget and time needed for the project realisation in Fel III up to +/−10%.

In budgeting the project, the Owner and his staff must be very careful. Time and again projects are budgeted incorrectly causing major cost and time overruns.

The major sources of risk are:

Inaccurate forecasts of project costs, project demand and other impacts.

Inaccuracies in management of the project in early stages because of bias, being optimism leading to falsely estimating cost and time, and strategic misrepresentation or, in plain English, lying about costs and time needed for completion.

4 The contract conditions in the General Conditions part regulate both the Contractor's and the Owner's obligations and rights and in order to enhance the achievement of the project objectives are best when they are balanced and do not favour one party over the other. Standard forms of contract conditions are usually the best particularly if they have been written with balance in mind. Examples of balanced contract conditions are NEC, JCT and ICC. The Special Conditions make the contract conditions project specific.

5 The parties to the contract should seriously consider including a mutual hold harmless provision in the contract with an obligation to have the insurers of the parties waive their rights to subrogation as this will prevent endless litigation taking place when an accident happens, particularly causing injury or death.

CHAPTER 4

The tendering for and negotiation of a contract

4.1 Introduction

The Employer's Requirements as described in the previous chapter form the documents that the Contractors will bid on. They specify the project's requirements in terms of performance parameters for the project. In an EPC contract they may contain a few technical specifications, but these should be limited by the Owner as mistakes in such technical specifications may cause the project to not achieve the performance parameters and it will be very difficult in most cases to unravel whether or not the Contractor could and should have seen the mistake in the technical specifications, particularly if he could have and should have noticed the mistake in the tendering stage. Besides, if one hires a qualified Contractor to do the design, engineering and construction, then why would one provide technical specifications oneself? To assert important factors? Then list those factors in the Employer's Requirements and use them to evaluate the Contractors' tenders.

4.2 The Owner

With the Employer's Requirements come the Owner's choice of the conditions of contract, standard form of contract conditions such as NEC, FIDIC, JCT and ICC just to name a few. The alternative to a standard form of contract conditions is a bespoke contract, put together by the Owner's lawyers, which will be scrutinised by the tendering Contractor's lawyers for provisions that are unfavourable to the Contractor most likely, or better: unfavourable to achieving the project's objectives because they are unbalanced and thus solely protect the Owner's interests but not the project's interests and that will subsequently compel the Contractors to increase their price as to have contingencies for when things may go wrong during the execution of the project. The best contract conditions are balanced contract conditions that allot the risks to the party that is best able to prevent the occurrence of the risk event or price the consequences of the risk when it cannot be avoided if it materialises. The Contractor cannot always prevent risks from occurring, and to be able to make an adequate estimate of the cost of the consequences and the time required to overcome them very often requires faculties that humans do not possess. This is why the use of the word "Turnkey" should not be used lightly and as a relief for the Owner of liability for all project risks as argued in the previous chapter. When the Employer's Requirements are put together as described earlier and the more complete they are with accompanying information about the site, site conditions, site ambient conditions, feedstock particulars, sketches, etc., the more risks will have been prevented from occurring by getting them out in the open, and then it is up to the Contractor to assess them and incorporate counter measures in his tender. This is the reason why it is not a great idea to award the contract for the project to the Contractor that has submitted the lowest price in his tender. It is true that the award criterion must be an objective criterion instead of a subjective criterion as tendering Contractors must be treated fairly by Owners both in public procurement but also in non-public procurement in most jurisdictions, but the lowest price is not the only objective criterion that exists. The Owner could attribute points to the way the Contractor has addressed the project's risks in his tender and the method of evaluation of the Contractor's tender in that respect, and subsequently award the project to the Contractor that scored the most points in addressing the project's risks.

The Owner must also choose the contract conditions he wishes to be part of the contract; they determine how the various risks will be dealt with during the course of the project. The Owner has various options: he may choose the most balanced standard conditions of contract, or the ones he feels "protect" him the best by allocating most risks to the Contractor, or draw up his own bespoke conditions of contract. As said the best contract is a balanced contract and the choice for a certain set of conditions should be well considered. External lawyers will see their job as protecting their Owner, as will many in-house lawyers, but guiding the decision maker to the right choice requires in-depth knowledge of the whole process, the technology and also the courage to divert from the general opinion within the Owner's organisation about which conditions to use for a particular project. Good leadership is necessary to allow opinions to be expressed, and wisdom is necessary to make the right choice.

If one is looking to achieve the project's objectives as an Owner, then collaboration is the best way, not just by mouth but by actions, and exactly the same goes for the Contractors. The construction industry must do away with the old antagonistic approach that has never worked, and people working in the industry must reset their minds to collaboration and focus on the project objectives instead of other objectives.

Award criteria should be the economically most advantageous tender, considering not just the construction cost but also the maintenance and operations cost for a given number of years and as said earlier the tenderers approach to the project risks and the tenderers commitment to collaboration with the Owner witnessed by reference letters from Owners in previous projects of similar size and functional aspects.

Once the FEED has been completed the Owner should not set his budget at −10%, or worse −15%, of the outcome of FEL III and then manipulate the bidding process in such a way that some bidding Contractor is dumb or desperate enough to agree on such an inadequate sum of money or even so devious that he counts on being able to change things around during the construction process.

Seventeenth-century English philosopher John Ruskin once said:

> It is unwise to pay too much, but it is worse to pay too little. When you pay too much, you lose a little money – that is all. When you pay too little, you sometimes lose everything, because the thing you bought was incapable of doing the thing it was bought to do. The common law of business balance prohibits paying a little and getting a lot – it can't be done. If you deal with the lowest bidder, it is well to add something for the risk you run. And if you do that, you will have enough to pay for something better.

Owners should bear this in mind!

4.2.1 The science of it all

Extensive scientific research has been done and described by Professor Bent Flyvbjerg of the University of Oxford. In his paper "From Nobel Prize to Project Management: Getting Risks Right", Professor Flyvbjerg explains what the major sources of risk in construction are:

1 Inaccurate forecasts of:

   a Project costs,
   b Project demand,
   c Other impacts.

2 Inaccuracies in Project Management:

   a Bias,
   b Strategic misrepresentation.

Bias, he explains, can be:

1 Psychological: optimism that things are the way you want them to be, so self-deception (happens in organisations where political and organisational pressures are low), or it can be
2 Political: strategic misrepresentation (happens in organisations where political and organisational pressures are high):

   a Overestimating benefits,
   b Underestimating costs.

Flyvbjerg sees as a solution: "Reference Class Forecasting" to beat the Planning Fallacy which entails that people tend to underestimate costs, completion times and risks of planned actions while at the same time they overestimate the benefits of those actions. It was developed by Kahneman and Tversky in their work on decision-making under uncertainty. They won the Nobel Prize in Economics in 2002. Their work shows that errors of judgment are often systematic and predictable rather than random, manifesting bias rather than confusion and thus corrective prescription must reflect that.

Reference Class Forecasting comprises three steps:

1 Identification of a relevant class of past similar projects, which class is to be broad enough to be statistically meaningful and narrow enough to be truly comparable;
2 Establishing a probability distribution for the selected reference class through credible empirical data to make a statistically meaningful conclusion;
3 Comparing a specific project with reference class distribution to establish the most likely outcome for the specific project.

1+2 + 3 will yield a statistical outcome that certainly can help Owners to decide which projects to pursue and which not to pursue, albeit preferably with the help of statisticians.

4.3 The Contractor

Tender preparation involves highly skilled personnel and considerable time for Contractors and is thus a costly activity for the Contractor with no guarantee that such cost will be recovered through the award of the prospective project. The Contractor can retrieve such cost by adding them to the calculation of the annual level of "overhead" cost, the not so categorisable and identifiable cost of running his business, but obviously that will not make him more competitive.

Contractors must and do strive constantly to get better at their trade, that is the only way for them to survive, but they need the execution of projects to be able to do that. And so, with every tender the first goal of the Contractor is to be invited by the Owner to discuss his tender and to negotiate for the contract, in other words to have a "fighting chance", and the predicament Contractors are in must not be abused by Owners to get a lower price than would be conducive to achieving the project's objectives. So, the Owner must accept that he will have to pay the Contractor the cost of the project and a fair margin on top of that cost. Who decides what a fair margin is on top of the Contractor's cost? Well, the Owners by deciding how much they are prepared to pay and the investors in construction companies, because otherwise no one will invest in construction companies and then soon we won't have any construction companies anymore. This is why Contractors continuously have to invest in improving their performance – more efficient, better quality, "cutting edge" and "at the top of the game".

When assessing the contract conditions that are part of the invitation to tender, the Contractors must qualify and want to renegotiate the following issues or run the risk of going bankrupt if a major risk materializes in the project:

4.3.1 Contractual exclusions*; in the contract conditions*

Exclude any liability for financial losses or damages such as, but not limited to: loss of profit, loss of production, loss of revenues, as well as for any consequential or indirect damages or losses.

Exclusion is to be irrespective of the possibility to recover the same from insurance or from third parties and whether or not a consequence of physical damage.

Exclusion must take the form of an indemnity to protect the Contractor against financial losses claimed by Owner's affiliated companies, co-venturers and other Contractors. These issues must be excluded in the contract because when they materialize they may cause bankruptcy for the Contractor. To not exclude them is self-protection by the Owner and not conducive to a successful project; these losses and consequential damages can be prevented by cooperating towards a successful project but not by the old-school antagonistic ways.

4.3.2 Contractual warranties/liabilities

- Contractual warranties/liabilities should be limited in time and as to available remedies.
- There should be no other warranties and liabilities than those expressly mentioned in the contract.
- The contract must have an overall liability cap, the size and content of the cap as well as the required carve-outs depend primarily on the industry – civil construction will be different than oil and gas.
- Performance liabilities should have the form of make-good obligations and liquidated damages as sole remedy.

The Contractor's management that must make the decision to tender for the project or not must have a complete understanding of the tender documents and what they require to be able to assess all the risks the company will take on if the project is awarded to them.

They need the help of their in-house specialists for the individual items and could be helped with regard to the legal and contractual risks from the contract provisions by a table put together to assess the said risks and discuss with the tender team what can be accepted and what not and what should be done to qualify the provisions that are not acceptable and how they should be amended.

The topics for such a table should be in column one, titled **Contractual Exclusions**, and it contains the Contractor's required contract provisions the so-called "must haves" in the Contract:

1. Exclusion of any liability for financial losses or damages such as – but not limited to – loss of profit, loss of production and loss of revenues, as well as for any consequential or indirect damages or losses.
2. Such exclusions are to be irrespective of the possibility to recover the same from insurance or from third parties and whether or not a

consequence of physical damage (***guidance note:*** *The exclusion should preferably take the form of an indemnity to protect the Contractor against financial losses claimed by Owner's affiliated companies, co-venturers and other Contractors*).

3 Warranties and liabilities are to be limited in time and as to available remedies. There can be no other warranties and liabilities than those expressly mentioned in the contract.

4 The contract must have an overall liability cap. The cap depends on contract type, size and content of possible exclusions. The maximum overall liability cap is to be X percent (X%) of the contract price. The overall cap must be kept as low as possible, whilst considering the value of the contract and any carve-outs.

5 Except for possible make good, liquidated damages to be the sole remedy, limited to a maximum percentage of X percent (X%) of the contract price and included in the overall liability cap.

6 Delay: liquidated damages as sole remedy, limited to a maximum percentage of X percent (X%) of the contract price and included in the overall liability cap.

7 Mechanical warranty triggered by actual hand-over (partial or complete).

8 Mutual hold harmless for loss and damage to property and death and injury of personnel between Contractor group (this group shall include Contractor's affiliated companies, subcontractors and vendors/suppliers) and Owner's group (this group shall include Owner's affiliated companies, agents, co-venturers and other Contractors), irrespective of negligence or breach of duty.

9 Third parties: contractor should obtain indemnity from the Owner for liability arising from damage caused to third parties not part of Owner's group in case of significant exposure on the work site. Failing such an exposure, or for third parties other than identified third parties concerned by such an exposure, Contractor may accept to be liable according to the applicable law (i.e. on an 'at law' basis), without indemnity.

10 Contractor must obtain indemnity from Owner for liability arising from pollution and contamination of any kind emanating from any spread, equipment or property (other than Contractor's spread, equipment or property).

11 No variation (or an aggregate of several variations) can be imposed on Contractor above a certain threshold (to be approved on a case by case basis).

12 The contract must provide a mechanism to mutually agree upon price, schedule and other changes needed to reflect the impacts of variations in a timely fashion.

13 Contractor must retain exclusive ownership on its proprietary information.

14 Owner's use of proprietary information must be limited to the specific contractual purpose (the "Permitted Purpose").
15 Contractor's proprietary information must be subject to secrecy obligations for at least X years.
16 There must be a mutual indemnity for patent infringement.
17 Disputes must be settled through internationally recognised arbitration rules (such as ICC, LCIA, UNCITRAL) preferably by an organisation that provides, next to rules, also a court that monitors the quality of the arbitration process such as ICC or LCIA or a national arbitration institute in the country where the site is located.

In the second column one puts the relevant provision of the contract conditions and in the third column what qualification will be proposed and negotiated with the Owner. This enables the internal discussion within the Contractor's management, the key persons in the project team and the sales/negotiating team very well. It is a systematic approach of preparing the Contractor's staff for the negotiations and over time they become very aware of the key points in contract conditions that weigh heavily for the Contractor.

Admittedly, in civil construction there are many competitors and it will be much harder to achieve all qualifications to be inserted in the contract; the oil and gas engineering Contractors have the luxury of having far less competitors and for them it may be easier to get their qualifications accepted by the Owner.

Negotiating contract conditions between the Contractor and the Owner needs to be aimed at creating the best environment to achieve the project's objectives, that it can be completed:

- In the available time;
- Within the available budget;
- Delivering the required outputs and performing as expected or better;
- Meeting the approval of all relevant stakeholders;
- Without any serious disputes.

Close cooperation between Owner and Contractor certainly can make this happen, but in the light of history, it will take courage, both on the side of the Owner and on the side of the Contractor, and courage is needed from the very beginning, but the reward is a good project. It is like approaching the vicious dog of a neighbour: perhaps you are approaching him the wrong way and that is why he growls and nips at you; be calm and confident and he will most likely be a different dog to you, so become the "Contractor whisperer" or the "Owner whisperer".

4.4 Contract conditions

As stated before, contract conditions must be balanced, allocating risks in such a way that risks are allocated to that party that is best able to either control a risk materialising or best able to price its effects when it materialises.

The conditions need to cover:

1. The definitions of the defined terms in the contract;
2. When the contract enters into force;
3. That the parties must act in accordance with the principles of Good Faith and Fair Dealing towards each other;
4. The ruling language of the contract and communications;
5. Applicable substantive law;
6. New or changed laws, standards and regulations, what to do if they affect the performance of the contract;
7. Interpretation of the contract, order of priority of contract documents;
8. Gaps in the documentation and information, what to do, who is responsible;
9. Obtaining permits, planning and other permissions;
10. Guarantees, bonds or other securities;
11. Demands under guarantees, bonds or other securities;
12. No agency relationship, parties independent from one another;
13. Contractor's general obligations;
14. The Owner's general obligations;
15. Coordination with other Contractors on site by the Owner;
16. Quality assurance;
17. Staff and labour;
18. Representatives of the parties;
19. Each party's duty to notify;
20. Scope of the contract price and works;
21. Setting out on the site;
22. Site information for the Contractor's design and use;
23. Unexpected artificial or physical conditions or obstructions;
24. Safety;
25. Public convenience;
26. Environmental protection;
27. Services or supplies to be supplied by the Owner or public utilities;
28. Owner-supplied plant, materials or equipment;
29. Responsibility for shipping and storing goods and materials;
30. Ownership of goods, materials, plant and equipment;
31. Keeping the site clear;
32. General allocation of responsibility of the parties for the design;
33. Design review process;
34. Intellectual property rights;
35. Variations and changes to the Works;
36. Start date, Time to Taking-Over, scheduling and progress;
37. Delay damages;
38. Schedule of contractual dates, time schedule and reporting progress;
39. Progress reports;

40 Contract price;
41 Payment of the contract price;
42 Applications and procedure for payment;
43 VAT/GST or equivalent consumption taxes;
44 Financing charges for late recognition of rights between the parties;
45 Completion;
46 Commissioning and performance tests and retest;
47 Taking-Over/provisional acceptance;
48 Assessment procedures after Taking-Over;
49 Defect Correction Period;
50 General principles of risk allocation and responsibility;
51 Liabilities of the parties and their reciprocal limitations and exclusions;
52 Fraud, wilful misconduct or gross negligence;
53 Warranty for future purchasers or Operators;
54 *Force Majeure*;
55 Suspension, termination and/or rescission of the contract;
56 Insurance;
57 Confidentiality;
58 Prohibition of corruption and compliance with applicable laws and standards of ethics;
59 Exclusive remedies, entire agreement, severability, amendments to the contract, waivers;
60 Joint and several liability;
61 Assignment of the contract;
62 Communications, including notices and consents;
63 Claim and dispute resolution procedures;
64 Arbitration.

The contract provisions decide the risk profile of the contract, and thus the estimation of the time for completion and the cost of the Works, the higher the risk profile the more time will be perceived necessary for completion and the higher the cost perception will be on the Contractor's side. This cannot be compensated for by awarding the contract for the lowest offered price for the reasons addressed earlier.

It has become obvious, not in the least through scientific research as done by Professor Bent Flyvbjerg and Professor Frank Emmert and others, that the old antagonistic approach of one versus the other by the Owner and the Contractor does not work, leads to not achieving the project's objectives and results in cost-overruns and delay of completion. The distrust in each other is tangible in every negotiation for a project. Those involved in the process shrug and accept that "it is what it is" and blame the cost-overruns and delays on the other which solidifies the status quo.

To avoid cost-overruns and delays each party's attitude must change towards a collaborative approach, which often does not happen because each of the parties thinks: "What if I do, but the other doesn't? I will not be tricked by the other(s)!" And thus, the distrust and blame game continues, to the detriment of the projects, the parties and in public projects those who ultimately pay for the projects, the tax payer.

It is of the greatest importance to leave the antagonistic attitudes behind in the negotiations before the award of the project. Both the Contractor and the Owner must instruct their negotiating teams to focus on achieving the project's objectives. This has nothing to do with agreeing to everything the other side says but is essential to explain why certain things are needed in the contract for the project to end well.

For the Owner

Price certainty through a fixed, lump sum price;

Time of completion certainty by fixed date for completion and losses due to delays compensated for by liquidated damages payable by the Contractor;

Certainty of performance as agreed by testing before acceptance and liquidated damages for not performing as agreed, secured by on-demand bank guarantees that can be drawn when tests prove the Works do not perform as agreed.

For the Contractor

Certainty that the Owner has the finances in place to pay for the contract price;

Payment as agreed upon achievement of milestones;

Not having to accept risks of an unknown or undisclosed nature;

Access to the site and ability to work undisturbed by others.

In lump sum contracts Contractors often refuse to give details of how their price was built up, in other words: to itemize the various components of their bid and price them individually amounting to a lump sum price. This is out of fear that such detail will be used against them in negotiations: "Bidder A quotes design at 4.5 million and you quote it as 5 million. How is that?" And if cost of materials increases during the execution of the Works and a variation is discussed they will be arguing with the Owner over the price offer for the variation order the Owner is contemplating. That may be a possibility, but given the fact that Owners are often very optimistic in their own price and time estimates of variation orders they are contemplating, giving some detail in the build-up of the price may actually help the Contractor in the attempt to come to an agreement on the variation proposal made to the Owner. See Section 5.4 in the next chapter.

When negotiating for the award of the contract the parties must depart from the project's objectives in everything that comes to the table, avoid trying to figure out why what the other side is saying is better for him, check what he is saying against the project's objectives and if what he is saying is detrimental to the project's objectives then explain why. So: do not bargain over positions, separate the people from the problem and insist on objective criteria[1].

Extensive scientific research has been done and described by Flyvbjerg. In his paper "From Nobel Prize to Project Management: Getting Risks Right" Professor Flyvbjerg explains what the major sources of risk in construction are:

1 Inaccurate forecasts of:

   a Project costs
   b Project demand
   c Other impacts;

2 Inaccuracies in Project Management:

   a Bias
   b Strategic misrepresentation.

Bias, he explains, can be:

3 Psychological: optimism that things are the way you want them to be, so self-deception (happens in organisations where political and organisational pressures are low), or it can be
4 Political: strategic misrepresentation (happens in organisations where political and organisational pressures are high):

   a Overestimating benefits
   b Underestimating costs.

Flyvbjerg sees as a solution: "Reference Class Forecasting" to beat the Planning Fallacy which entails that people tend to underestimate costs, completion times and risks of planned actions while at the same time they overestimate the benefits of those actions. It was developed by Kahneman and Tversky in their work on decision-making under uncertainty. They won the Nobel Prize in Economics in 2002. Their work shows that errors of judgment are often systematic and predictable rather than random, manifesting bias rather than confusion and thus corrective prescription must reflect that.

Reference Class Forecasting comprises three steps:

1 Identification of a relevant class of past similar projects, which class is to be broad enough to be statistically meaningful and narrow enough to be truly comparable;
2 Establishing a probability distribution for the selected reference class through credible empirical data to make a statistically meaningful conclusion;

1 Roger Fisher & William Ury: *Getting to Yes* (Random House Business Books, London, 2011).

3 Comparing a specific project with reference class distribution to establish the most likely outcome for the specific project.

1+2 + 3 will yield a statistical outcome that certainly can help Owners to decide which projects to pursue and which not to pursue.

4.5 Takeaways

1 Choosing the contract conditions and thereby the balance, or lack thereof, between the parties' obligations and rights are normally the Owner's choice. A choice to allocate all the risks towards the Contractor does not bode well for the achievement of the project objectives. Owners should not subscribe to a disturbing trend by allocating "all the risks" to the Contractor by giving the contract the label: "Turnkey". That is not what the word Turnkey is about, it is about the most extensive scope of work for the Contractor, and such an extensive scope brings with it more risks but not all risks. An attempt to transfer all the risks to the Contractor would mean that the project would most likely become unaffordable for any Owner.

2 In budgeting the project, the Owner and his staff should be very careful. Time and again projects are budgeted incorrectly causing major cost and time overruns.
The major sources of risk are:

   a Inaccurate forecasts of project costs, project demand and other impacts
   b Inaccuracies in management of the project in early stages because of bias, being optimism leading to miscalculation of prices and time for completion and strategic misrepresentation or, in plain English, lying about costs and time needed for completion.

3 Reference forecasting seems to be the way to beat the Planning Fallacy, which entails that people tend to underestimate costs, completion times and risks of planned actions while at the same time they overestimate the benefits of those actions. It is a three-step process:

   a Identification of a relevant class of past similar projects, which class is to be broad enough to be statistically meaningful and narrow enough to be truly comparable
   b Establishing a probability distribution for the selected reference class through credible empirical data to make a statistically meaningful conclusion
   c Comparing a specific project with reference class distribution to establish the most likely outcome for the specific project.

These three steps when executed well will yield a statistical outcome that certainly can help Owners to decide which projects to pursue and which not to pursue.

CHAPTER 5

Basic Design and Engineering

5.1 Introduction

This chapter goes into the intricacies of the first stage of design: the Basic Design and the engineering process.

When a FEED has been done, there already is much of a Basic Design. If the Contractor who was awarded the contract for the project has not done the FEED himself, design liability for the Basic Design remains with the Owner. If the FEED was done by the Contractor who is now contracted to execute the project design liability rests with the Contractor. If a different Contractor is selected to execute the project, the design liability will remain with the Owner, who will have recourse on the Contractor who executed the FEED of course. It therefore makes sense to always check the Contractor's liability insurance by asking for an insurance certificate from his insurers regarding the work the Contractor will be doing, before awarding him the contract for the work.

This chapter continues on the basis that there was no FEED done and it is within the Contractor's scope of work to execute the Basic Design on the basis of the Employer's Requirements.

5.2 Design liability

Using the Employer's Requirements and accompanying information, the Contractor is to execute the Basic Design. Considering the size of the available plot of land on the site, the Contractor must fit the project into the available space. Considering the site data and subsoil conditions he must design the necessary foundations, etc., all of which he already must have considered when preparing his bid for the project. The Basic Design forms the basis for the detailed engineering of the entire project. Should the Contractor discover any mistakes, omissions or faulty information he must warn the Owner thereof forthwith. In many jurisdictions it would be a legal obligation to do so but not in every jurisdiction, so it is advisable to put this obligation in the contract as a contractual obligation.

Design, in general, is an iterative process of refining the engineered scope of a project[1]. It goes from high-level to the definitive design in a series of mile-

1 Jonathan Hosie: *EPC Articles* (Mayer Brown, London, 2016).

stones: packages for piling and foundations; Piping & Instrumentation Diagrams (P&IDs); detailed drawings for procurement and construction; and Issued For Construction (IFC) drawings for an industrial plant in oil and gas.

In civil engineering it is often a backwards process; once the Basic Design of superstructure of the bridge is designed the engineers can calculate the foundations of the bridge. This is because they need to know the "load" on the foundation and the subsoil composition and stability for the design of the foundations.

During the design process it is essential that the Contractor notifies the Owner of any problems encountered in the Employer's Requirements and accompanying information that will have an effect on the cost of the project and or will cause a delay in the completion.

As said before: the Employer's Requirements form the foundation of the project, if the foundation of a building is compromised, no one will be surprised if this will result in damage to, or even collapse of, the building. So why would anyone be surprised if faults in the Employer's Requirements lead to extra cost and need for extra time?

The various engineering disciplines within the Contractor's organisation need to regularly have contact with each other regarding the design work for a project. After all the designs are brought together everything must fit. More than once the author has witnessed a process engineer and a piping engineer discuss a design at the coffee machine.

> Piping engineer: "I am changing the piping lay-out because it will be easier to construct," and the process engineer responded: "No you won't because if you do that the process will not work and the plant will not produce what we contracted with the Owner."

Needless to say, such discussions should not take place at the coffee machine but in the project meeting. The fact that engineers of different disciplines talk to each other is good, but it is better to have it happen in a meeting with the whole of the project team present.

The Contractor must warrant that it, its designers and design subcontractors have the experience and capability necessary for the design. The Contractor must undertake that the designers used for the Works will be available to attend discussions with the Employer at all reasonable times until the design phase is completed.

When design documents are completed they must be submitted to the Owner for review. As the Owner pays for the design work, he has an interest in seeing that the design work is progressing and progressing in an orderly fashion. The contract should indicate which design documents are to be submitted to the Owner and the Owner's review should have a (relatively short, e.g. 21-day) time limit in the interest of progress of the design works. When the time limit is exceeded by the Owner the relevant design documents are to be deemed approved, so that the Contractor may continue the design work. The Owner returns the design documents marked: "Reviewed by the Owner/Employer" and dated. If the Owner is of the opinion that a design document does not comply with the contract he must notify the Contractor thereof and the Contractor must rectify the document and resubmit it to the Owner. Acceptance by the Owner of design documents does not transfer design liability from the Contractor to the Owner.

The Contractor will only be discharged of his design liability when the final tests prove that the Works perform as contracted and the Certificate of Completion has been handed to the Contractor and the Works have been accepted and taken over by the Owner.

Even if the design is compliant with the contract, the Owner must have the right to instruct changes to the design which instruction is then subject to the "variations" provision of the contract; however the further the design work has progressed the more costly changes tend to become because earlier design work may also have to be changed to "fit" the changed design.

It may also be that the Contractor wishes to make changes to the design documents by proposing to the Owner a variation in which the Contractor must illustrate how the proposed change will improve the quality, efficiency or safety of the Works, or will reduce the expense to the Employer of executing, maintaining or operating the Works. The Contractor must outline the benefits of any proposed variation he perceives, including the financial aspects. The value of the benefits, unless otherwise agreed, should be shared in an agreed proportion between the Contractor and the Owner.

5.3 The complexities of design

As design is often spread over various engineering disciplines, particularly in the oil and gas industry, it is done in different departments and the engineers need to be in regular contact with each other as in the end all the various design efforts must "fit". That may seem obvious but in practice it isn't always, particularly if the various design activities are executed in a Consortium of various Contractors and the engineers are not all located in the same building. This may seem too obvious to mention but as it sometimes does lead to problems, it seemed worth mentioning.

The various design elements must fit; if one department designs a turbine compressor package, it must fit in the space designated for it in the building in which it will be installed, which will be designed by a civil engineering department. Communication therefore is of the essence.

Design development is one of the key risks under EPC contracts, also possibly for the Owner if he wants to maintain control over the process and its progress.

5.4 The variations provision

If the Owner wishes to implement a variation to the Works, the Owner must send a notice to the Contractor describing the variation. In that notice he should also request that the Contractor provide a reasonable estimate of cost of and time required for preparing a variation proposal. The contract must stipulate how much time the Contractor will have to prepare such a proposal and the Owner should be aware that some proposals take more time than others to prepare due to the extent of the variation and the effect if may have on work already completed. Also, it is counter-productive to achieving the project objectives when the Owner frequently requests that the Contractor provide him with variation proposals. It seriously disturbs the design work being done and thus causes delays and increases cost for the Contractor.

The Contractor must provide the Owner with a proposal comprising:

1 A description of the proposed design and work to be executed;
2 A program for the execution of the variation;
3 A proposal for any necessary modifications to the Time Schedule and the Time to Taking-Over;
4 A price for the variation which can either be lump sum or a price estimate comprising quantities and either a list of all-in unit prices or a list of prices per unit subject to the application of a mark-up for overheads and profit if such list is part of the Special Conditions. The Owner should be able to audit the application of indexes (if any) to the adjustments of the prices of such units;
5 If the variation is proposed by the Contractor: a description of the improvements the Contractor expects the variation to result in, including where applicable estimates of the reduced expenses and proposed allocation of the value between the Owner and the Contractor, and
6 Other elements of the Project likely to be affected by the variation.

Subsequently the Owner must be granted a limited time from the receipt of the Contractor's proposal to decide whether or not the variation will be implemented. The Contractor must be notified forthwith of the decision by the Owner. If the Owner does not accept the Contractor's proposal for the variation, yet instructs him to proceed with the variation he must indicate in the notification the reasons for not accepting the Contractor's proposal and how the Contractor will be compensated for the work and how the time schedule is to be adapted. If the Contractor does not agree with the Owner's amendments to the proposed cost and time consequences, his only option would be to refer the dispute to dispute resolution in accordance with the contract, depending on the contract conditions, of course. He also might want to weigh the cost of such a step against the difference between his assessment of the cost and the Owner's assessment of what he is prepared to pay for the execution of the variation.

5.5 Takeaways

1 When the Owner has a Front-End Engineering and Design (FEED) done, he will have a good idea what the cost, the time for completion and the design of the project will be. If the Contractor who executed

the FEED is selected to execute the project, the design liability will remain fully with him. If a different Contractor is selected the design liability will remain with the Owner, who will have recourse on the Contractor who executed the FEED, of course. It therefore makes sense to always check the Contractor's liability insurance by asking for an insurance certificate from his insurers regarding the work the Contractor will be doing.

2 Using the Employer's Requirements and accompanying information, the Contractor is to execute the Basic Design. Design is a reiterative process that is based on the Employer's Requirements and the accompanying information provided by the Owner. Imagine a long row of domino bricks; when one falls it is likely that the rest will also fall, meaning a change in one of the phases of design may have consequences for both subsequent design and also previously completed part(s) of the design. So, faults in the Employer's Requirements and incorrect information may have very significant consequences both in cost and in time for completion.

3 It is important that the various design disciplines have regular and organised contact with each other, preferably in the presence of the Project Manager and the Contract Manager. This enables them to share information on issues that are different than expected from the Employer's Requirements and allow them to adapt their work directly after discovery and the Project Manager and the Contract Manager can establish who is to pay for any extra cost and if there is an entitlement for extra time for completion.

4 The Owner must approve design within the time limits given by the contract. Approval of design does not transfer responsibility and liability for the design to the Owner. Under an EPC/Turnkey contract the Contractor must deliver the Works in accordance with the functional specifications. Any design approval by the Owner does not change that.

5 The variation provision of the contract must be followed meticulously by the Contractor when the Owner wants a change to the Works. The Contractor must prepare a proposal comprising

a A description of the proposed design and work to be executed;
b A program for the execution of the variation;
c A proposal for any necessary modifications to the Time Schedule and the Time to Taking-Over;
d A price for the variation which can either be lump sum or a price estimate comprising quantities and either a list of all-in unit prices or a list of prices per unit subject to the application of a mark-up for overheads and profit if such list is part of the Special Conditions. The Owner should be able to audit the application of indexes (if any) to the adjustments of the prices of such units;

e If the variation is proposed by the Contractor: a description of the improvements the Contractor expects the variation to result in, including where applicable estimates of the reduced expenses and proposed allocation of the value between the Owner and the Contractor, and

f Other elements of the contract likely to be affected by the variation.

6 The Owner must refrain from regularly asking for variation proposals from the Contractor, as the work involved in the preparation of such proposals is very disturbing for the other work on the project.

CHAPTER 6

The procurement of services and materials

6.1 Introduction

This chapter will go into the (contractual) relationship that the Contractor will enter into with other companies that will be engaged to execute part of the (Main) Contractor's scope of work and the suppliers of materials and equipment, necessary for the execution of the Works under the main contract.

6.2 The subcontract conditions

The Contractor himself will procure both services and materials for the execution of the project when the project is of such significant size that he will not execute all of the Works himself. Very often they procure services and materials under contract conditions they have prepared themselves to protect their own interests. Subcontract conditions under which they wish to procure the services of a subcontractor are such that they would never consider entering into a contract with an Owner under such conditions. These are the conditions that shift all the risks of the scope of work under the subcontract to the subcontractor in the same way that Owners often try to do to them and which they will never accept. Then why would one try to force such contract conditions onto another Contractor who is to execute part of the scope of works under the main contract for the (Main) Contractor? It is very likely that the subcontractor will be inclined to refuse such conditions and if he is not in a position to refuse, to try and "outsmart" the Contractor during the execution of the subcontract works. All this is not helpful in the achievement of the project objectives.

In fact, under a subcontract the scope of work is part of the scope of work of the main contract between the Contractor and the Owner and the question then is: Why would the contract conditions under which the subcontractor must execute his work be vastly different? As said before: "The best contract is a balanced contract", and that also applies to subcontracts. A cooperative way of working is a better guarantee to achieve the project's objectives than an antagonistic way.

So, if the subcontract scope of work is part of the scope of work of the main contract then the subcontract conditions must also be balanced in order to achieve the project objectives (remember: *without disputes*).

A good way of achieving that would be to use the conditions of contract of the main contract, albeit that the subcontract specific issues such as the scope of work and the purpose of the subcontract works must be adapted as well as certain features of the main contract where the main contract gives the Contractor a given number of days to notify the Owner of issues that will affect the cost of the main contract works and the time for completion of the project. If the subcontractor gets the same amount of days for notification of such issues as the Main Contractor and he notifies the Main Contractor on the last day, then this will have detrimental consequences for the Main Contractor as he will be too late in passing on the notification to the Owner and depending on the conditions of the main contract perhaps lose his right to claim.

Specific topics that need attention are inter alia:

- Scope of work;
- Performance specifications;
- Price;
- Time for completion;
- Any and all provisions that limit the number of days for the effectiveness of any situation or the required period of notification of any situation of action to be taken under the (sub)contract, e.g. performance testing to commence, submission of invoices, submission of progress reports;
- Health, safety and environmental measures;
- Cap on liability.

Obviously, the liability exposure of the subcontractor cannot be equal to the liability exposure of the Main Contractor; his subcontract price will be a part of the main contract and therefore the maximum aggregate liability of the subcontractor must be proportionate to the subcontract price.

6.3 General purchase conditions

Many companies, world-wide, that buy materials on a regular basis for their business do so under general purchase conditions. These general purchase conditions aim to protect the buyer. Most likely the seller will only sell them under his sales conditions and there we have it, the battle of the forms: Who can cancel out the other's conditions successfully so that only his will be applicable? Hopefully the reader has understood it at this point: this is not helpful in the achievement of the project objectives. Therefore, it is better to make a tailor-made set of purchase conditions that are balanced and attuned to the achievement of the project objectives.

6.4 Takeaways

1 As the scope of work under a subcontract is part of the scope of work under the main contract which is to be integrated into the project works,

the subcontract conditions are best balanced in the same way as those of the main contract. Attention must be given to notification obligations in terms of days the subcontractor will have to send a notification, as to not compromise the Main Contractor's notification obligation towards the Owner.

2 The same can be said for the purchase conditions regarding materials that are to be bought for the realisation of the project; they as well are best attuned to the conditions of the main contract. Main Contractor and vendor must avoid the "battle of the forms" as such battle is not conducive to the achievement of the project objectives.

CHAPTER 7

Construction

7.1 Introduction

This chapter will highlight the particulars of the construction phase of the project, after the engineering is finished, the ready for construction drawings have been approved internally as well as by the Owner and the construction works can start, including the work by subcontractors and suppliers of materials and the like.

7.2 Construction management

A central role in the construction phase is for Construction Management: where the Project Manager oversees and leads the whole project execution the construction manager manages the construction work and plays a significant role within the project team, as in this phase physical elements are being made and put together and any issue that arises must be brought to his attention immediately and he is to coordinate the investigation into the cause and the effect of what is going wrong.

Key people in the project team in this phase are:

- The planning engineer;
- The financial controller;
- The Health, Safety and Environment ("HSE") manager;
- The Contract Manager.

Time management (making sure that the Works progress in accordance with the time schedule) and cost control (making sure that the Works remain within the budget and that payments for work done are received on time) must be the primary focus of the planning engineer and the cost controller. Particularly when the Works do not progress as planned, the construction manager must be alerted by the planning engineer and he must investigate why and how parts of the Works are being delayed. The planning engineer must be able to tell him where things are going wrong and what he thinks is the cause of the delay. Experienced planning engineers usually know a lot about the daily goings-on and would be able to give a relevant analysis.

The Health, Safety and Environment manager also plays a crucial role. Having been involved concerning HSE issues from the Tender Stage on and having been the one to put together the HSE plan for the project, he is to make sure that all personnel of the Contractor and the subcontractors and anyone else the Contractor brings on the site adhere to the HSE plan. Preferably, he has been involved in checking the subcontractor's safety performance while subcontractors are bidding to the Contractor and he has made a classification of the subcontractors regarding their safety performances.

Some may feel that an HSE manager on the project is just unnecessary cost. They are not, for when accidents happen, one finds out what the consequences are, not only for the injured person(s) and for their families but in fact also for their colleagues on the job. Also, with regard to serious accidents the authorities may step in, resulting in major delays and cost increases.

The project manager ties all the various construction activities together; maintaining an organised site (efficiency and safety), his primary responsibilities are:

- Managing contracts and obtaining required permits and licenses.
- Preparing and submitting construction project budget estimates and monitoring the established budget throughout the project:
    - Managing revenues/profits and implementing monthly and annual business plans;
    - Controlling work-related costs in order to stick to the established budget; and
    - Verifying and approving subcontractors' invoices, as well as managing subordinates' payrolls.
- Establishing working schedules and objectives for the entire project and supervising the overall progress:
    - Developing and maintaining detailed project schedules, including all phases and dependencies, such as design, material procurement and installation, as well as administrative tasks for multiple sites;
    - Supervising the project to ensure that it is progressing as expected, reporting its status to the Owner on a regular basis;
    - Conducting regular update meetings with team members to measure the success of each project;
    - Managing all installation and readjustment work, along with the personnel and its assigned tasks.
- Building, managing, and deploying a high-quality team:
    - Employing staff members and overseeing their activities, choosing the most skilled employees to trust them with specific duties or issues of the project;
    - Ensuring all team members are following and are trained on current health and safety policies and procedures;

  - Training the team on any field that could be considered necessary;
  - Making sure all expectations are met, both collective and individual, and resolving any disputes or complaints, whenever possible; and
  - Drafting the contractual agreements with architects, consultants, Owners, suppliers and subcontractors, and making any necessary changes.

- Providing an excellent service in order to ensure Owner satisfaction throughout the whole project with property Owners, corporate Owners and insurance companies:

  - Inspecting progress and work quality during and after the construction project.

- Acting as the company's representative to negotiate union contracts and business services matters:

  - Promoting a culture of "employer of choice" to attract, retain and motivate high-quality workers; and
  - Developing and maintaining strong relationships and acting as a primary liaison with policyholders, estimators, adjusters, etc.

- Purchasing the required materials and equipment, as well as construction lots:

  - Managing purchase orders, changing orders and chargebacks.

- Ensuring that all construction and safety codes are duly followed:

  - Being aware of potential construction hazards.

Daily tasks of the project manager

- Monitoring and reporting the progress and development of all construction projects, ensuring requirements and deadlines are met;
- Updating and regulating projects' budgets by managing revenues and profits;
- Visiting construction sites to ensure all subcontractors and subordinate staff are working according to provided instructions;
- Inspecting and reviewing every aspect of the project to ensure that all construction and safety codes, as well as all applicable regulations, are being followed;
- Directing and supervising workers.

7.3 Work at the site starts

7.3.1 Setting out

The Contractor begins by setting out the Works in relation to original points, lines and levels of reference given by the Owner. If no such points, lines and levels of reference have been given to the Contractor, this should be done as soon as possible to allow the Contractor to continue the Works.

Errors in setting out or failure to provide accurate set-out data that affect the performance of the Works must be rectified. Such errors if attributable to the Owner are to be rectified at the Owner's expense and the Contractor will be, if delayed or put to additional expense, entitled to an extension of time and/or payment as a variation. The Contractor must rectify, at its own expense, any error attributable to the Contractor in the positions, levels, dimensions or alignment of the Works.

7.3.2 Fossils, mineral deposits that are of economic interest and items of archaeological interest

Following the ground-breaking ceremony, the Works at the site commence. The Contractor will prepare for the laying of the foundations by digging out the soil to the required depth. During these excavations the Contractor may encounter fossils, mineral deposits that are of economic interest and items of archaeological interest. This must be notified to the Owner forthwith who must deal with such items in accordance with the laws of the country where the site is located. The Contractor must make sure that none of his personnel or others remove any of these items from the site. Should the finding of such items delay the Works and/or increase the costs, such a situation warrants a variation order in accordance with the contract.

7.3.3 Safety

Where the Contractor takes possession of sections of the site, he has the primary responsibility for safety in those sections. Accordingly, the Contractor must:

- Comply with all applicable safety regulations and his own safety plan.
- Take care of the safety of all persons entitled to be on site, and not knowingly create hazards for persons not entitled to be on the site;
- Use reasonable efforts to keep the site and Works clear of unnecessary obstructions so as to avoid danger to all persons;
- Provide reasonable fencing, lighting, and security of the Works until Taking-Over; and
- Provide any Temporary Works (including roadways, footways, guards and fences) that may be necessary, because of the execution of the Works, for the use and protection of the public and of Owners and occupiers of areas adjacent to the Site.

The Contractor must keep records concerning health, safety and incidents and such records shall be made available to the Owner at its request. Details of any accident are to be sent to the Employer as soon as reasonably possible after its occurrence.

As a part of his obligation to abide by all safety requirements and regulations in force at the site or in relation to the Works the Contractor must communicate all such requirements and regulations to the Contractor's employees and subcontractor(s) and instruct them to observe and abide by such requirements and regulations. The Health, Safety and Environment manager must set up training sessions to inform the Contractor's personnel and subcontractor's personnel of the specific health and safety issues and regulations regarding the site where they will be working.

The HSE manager must make sure that when all subcontracts and purchase orders are being prepared, they include terms requiring that all applicable safety requirements and security regulations for the site and the Works are to be complied with, and that the terms of all such requirements and regulations are available on request.

7.3.4 Public convenience

The Contractor must make sure that the execution of the Works will not interfere unnecessarily or improperly with the convenience of the public, or the access to and use of roads and footpaths, whether or not they are public or in the possession of the Employer or of others.

7.3.5 Environmental protection

Both Contractor and Owner must take reasonable measures to not damage the environment (either on or off the site) or cause a nuisance to people or property from pollution, noise or other results of its operations. Should such problems seem likely to arise, the parties shall seek guidance from the local authorities concerned.

The Contractor must make sure that air emissions, noise, surface discharges and pollution from the construction activities do not exceed the values indicated in:

- The contract (if any),
- The relevant permits,
- Third party licenses or authority approvals,

  and do not in any event exceed the values prescribed by applicable laws.

The Contractor cannot be responsible for any damage, nuisance or other effects to people or property resulting from pollution, noise or any other consequence of pre-existing conditions of the site, or the activities of other Contractors of the Employer or third parties.

This also entails that each party must store and handle hazardous materials it needs for the Works with care and diligence and in compliance with the applicable law and Good Practice. As between the Owner and the Contractor,

1 Contractor is to be responsible for the removal, remediation or other appropriate action with respect to any hazardous materials which:

a Were brought on to the site by the Contractor or any subcontractor and which are not needed for the Works; or

b Otherwise got to the site as the result of any negligent, deliberate or unlawful act or omission of the Contractor or any subcontractor.

2 Employer is to be responsible for the removal, remediation or other appropriate action with respect to any hazardous materials present at, on, in or under, or migrating and/or emanating to or from the site that are required to be removed or remedied by applicable law.

7.4 Services or supplies to be supplied by the Owner or public utilities

Unless specifically agreed otherwise, the Owner is to be responsible for the provision of all power, water and other services or supplies to the site. Unless agreed otherwise, the Contractor is to be entitled to recover increases in the costs of such supplies and services due to inflation of the currency of the country occurring after the reference date named in the contract.

If there are services to be supplied by the Owner, the Contractor must pay the Owner at cost.

Should the Contractor be delayed or suffer extra cost in its execution of the Works due to a failure or inadequacy in the supply of services by the Owner or services provided by any other service supplier of the Owner, the additional cost and any delay are to be treated as a variation.

7.5 Owner-supplied plant, materials and/or equipment

The Owner must provide the items of all goods, materials, plant and equipment, buildings and storage facilities that he agreed to provide in the contract. The Owner must, at his risk and cost, transport such items to the Contractor or the site, as agreed. Owner-supplied goods, materials or plant and equipment are to be offered as being fit for the purposes indicated in the Employer's Requirements and shall be insured by the Owner against damage or loss.

The Contractor is to visually inspect items upon receipt and must notify the Owner of any shortage, defect or default; then, either the Employer must rectify any shortage, defect or default, within reasonable time, or the Contractor, in agreement with the Owner, will carry out such rectification as a variation. After visual inspection, such items will be under the care, custody and control of the Contractor. The Contractor's obligations of inspection, care, custody and control do not relieve the Owner of liability for any defect or default which could not be identified through visual inspection.

The operation of plant and equipment must be in accordance with Good Practice and in accordance with specific details accompanying such plant and equipment.

7.6 Responsibility for shipping and storing goods and materials

Unless otherwise agreed, the Contractor is to be responsible for packing, loading, transporting, receiving, unloading, storing and protecting all goods and other things required for the Works and to be supplied by the Contractor, on his behalf, and the Contractor is to bear all expenses associated with these activities.

7.7 Ownership of goods, materials, plant and equipment

Unless otherwise agreed all goods, materials, plant and equipment supplied under the contract or on behalf of the Contractor should be owned by the Contractor until they are paid for or until Mechanical Completion whichever will be the earlier. On Mechanical Completion or payment, the title to the goods, materials, plant or equipment is to transfer to the Owner. Temporary Works and Contractor's equipment are to remain the property of the Contractor.

The Contractor is to warrant good title to all goods, materials, plant and equipment provided by it, and Contractor must warrant that title and ownership will pass and vest in Owner, free and clear, of any and all liens, claims, charges, security interests, encumbrances and rights of third parties. The Contractor must indemnify the Owner against all claims resulting from any breach of warranty.

7.8 Takeaways

1 A central role in the construction phase is for Construction Management: where the Project Manager oversees and leads the whole project execution the construction manager manages the construction work and plays a significant role within the project team, as in this phase physical elements are being made and put together and any issue that arises must be brought to his attention immediately and he is to coordinate the investigation into the cause and the effect of what is going wrong.
Key people in the project team, next to the project manager, in this phase are:

- The planning engineer;
- The financial controller;
- The Health, Safety and Environment ("HSE") manager;
- The Contract Manager.

Time management (making sure that the Works progress in accordance with the time schedule) and cost control (making sure that the Works remain within the budget and that payments for work done are received on time) must be the primary focus of the planning engineer and the cost controller.

2 The Health, Safety and Environment manager also plays a crucial role. Having been involved concerning HSE issues from the Tender Stage on and having been the one to put together the HSE plan for the project, he is to make sure that all personnel of the Contractor and the

subcontractors and anyone else the Contractor brings on the site adhere to the HSE plan. An HSE plan and HSE manager on the project are not a luxury; accidents and their aftermath are very disruptive and costly.

3 The construction manager together with the Contract Manager play a central role. The Works start by setting out the Works in relation to original points, lines and levels of reference given by the Owner.

4 During the excavations the Contractor may encounter fossils, mineral deposits that are of economic interest and items of archaeological interest. This must be notified forthwith to the Owner who must deal with such items in accordance with the laws of the country where the site is located.

5 Contractor must make sure that executing the Works does not cause disturbances to the public and the environment and that the execution is in compliance with applicable local rules.

6 Owner is to be responsible for supply of plant, equipment and materials that he agreed to supply and therefore also for the consequences if he fails to do so in time.

7 All goods, materials, plant and equipment supplied under the contract or on behalf of the Contractor should be owned by the Contractor until they are paid for or until Mechanical Completion whichever will be the earlier. On Mechanical Completion or payment, the title to the goods, materials, plant or equipment is to transfer to the Owner. Temporary Works and Contractor's equipment are to remain the property of the Contractor.

CHAPTER 8

Delays

8.1 Introduction

This chapter looks into the causes and the effects of delay in completion of activities to be executed under the contract. Many claims in the construction industry are about delay and its consequences. Delay causes losses for the Owner but also for the Contractor.

For the Owner the loss is caused by the fact that he can start using the completed works later than planned and therefore loses income and profit.

For the Contractor delay causes loss because he cannot execute the Works as planned and has his personnel and that of his subcontractors standing idle while waiting for the opportunity to start or continue their work.

When a delay occurs, recognisable as when the completion of an activity does not concur with the planned completion date on the project schedule but is later, the following questions arise:

- What caused the completion of this event to be later than planned?
- What will the effects of the delay be?
    - Will it delay the final Taking-Over of the Works by the Owner?
    - Will there be an increase in the costs of completion due to this event?
    - For whose risk and account is this event that caused the delay?

The questions may seem simple, but finding the answers usually is not.

When first noticing delay it is of paramount importance to immediately investigate what event has caused or is causing the delay and to preserve any relating documentation. What has caused the delay (which event): Why are we slower in progress than planned and who has failed to do what he is supposed to do or who has done something he should not have done that has caused the slowdown? Was the event our responsibility as the Contractor or that of the Owner, or vice versa as the case may be.

8.2 Delay-causing events

Delay-causing events can be categorised as follows:

- Strikes,
- Failure to grant access,
- Shortage of labour,
- Project Manager's or Owner's instruction,
- Late provision of information,
- Late approval of design documents,
- etc.

A causing event is an event that causes a delay to progress of an activity on the project schedule and that has an effect on the completion date because it will delay subsequent interdependent activities on the project schedule that will ultimately cause delay of the final completion. In other words: an event that causes delay to an activity on the critical path, an activity that determines the duration of the project, is relevant because the delayed activity is on the critical path and will extend the duration of the project.

An activity that is not on the critical path is not relevant when it is slower than scheduled as it has no effect on the scheduled duration of the project; such an activity is labelled "disrupted" when it gets behind in the project schedule but is not considered a delayed activity as it will not delay final completion.

A delay to completion only occurs when the scheduled date for completion has passed and it was caused by a delay to progress preceding the date for completion.

So when:

1 Delay to progress has happened,
2 The delay to progress has affected an activity on the critical path, and
3 The delay to progress has not been recovered before the scheduled completion date arrives and completion is in fact delayed,

there is a delay to completion.

Depending on who is responsible for the event causing the delay, the result will be either:

1 An extension of time for completion and possibly reimbursement of Contractor's time related costs, if the event was the liability of the Owner, or
2 Entitlement for the Owner to liquidated damages, or unliquidated damages for the period of delay of the completion date, depending on the contract, if the event was the liability of the Contractor.

8.3 Concurrent delay

Imagine two delay-causing events, one the liability of the Owner, the other the liability of the Contractor. They may occur at the same time, or they may occur one after the other. If they occur at the same time they are concurrent events which cause the same effect. If they occur one after the other it must be established if the effect of both events is concurrent or not.

If the two causing events delay either the progress of one activity, or the progress of more than one activity, but the delay occurs over the same time period and is felt at the same time, the delay is a concurrent delay.

Where the affected activities are on the critical path and cause the same delay to completion meaning the one without the other, and vice versa, would have caused the delay to completion, they affect the completion date to the same extent.

An example: a Contractor is selected in a bidding competition to engineer, procure and construct a chemical facility at a refinery expansion. When according to the schedule the time has come to start work on the site, the Contractor is late in mobilizing equipment, materials and his construction crew. The Owner sends a notice to the Contractor ascertaining the tardiness in mobilisation. The Contractor's representative that visits the site upon receipt of the notice to explain the delay sees that the site is flooded with seawater and contractually it is the Owner's responsibility to keep the site dry at all times after award of the contract to avoid salty deposits corroding the steel materials to be stored at the site. The schedule contains a tight planning with little time to spare. Both events, the delay in mobilisation as well as the flooded site, will delay the completion date of the project. Who is to be liable for the delay? Obviously both the Contractor and the Owner must send the other a notice of the other being in default under the contract. The Contractor replies to the Owner's notice that he has been ready to mobilize equipment, personnel and materials forthwith, but the site being flooded with seawater such mobilisation was impractical as work nor storage is possible at the site until the seawater has been drained from the site and all salty deposits have been removed. The Owner answers that the draining of seawater and flushing out of salty deposits will be done in one day. If the Owner can indeed drain the seawater and flush out the salty deposits in one day and the Contractor can mobilize immediately thereafter and start the site activities, but the date of completion of the project is delayed by all this, who is liable for the consequences of this delay? Is it the Owner – does he have to pay the Contractor for cost of equipment, materials and personnel being idle or is the Contractor liable for the contractual liquidated damages because he was in default mobilizing equipment, materials and personnel? The fact that the Contractor only signals that the flooded site makes late mobilisation an irrelevant issue in reply to the Owner's notice to the Contractor of late mobilisation makes the Contractor's point of view at least suspect. A prudent Contractor would immediately have sent a notice of the fact he did not have access to the site. As he did not, he makes himself a suspect of using the situation in his favour.

8.4 Sequential delay

If the Contractor is delayed during a period due to his own fault and at a later stage delayed again through an event that is at the Owner's risk but neither delay is on the critical path and does not cause delay of completion this leads to the following result:

- Contractor's right to an extension of time is not relevant, the completion date is not affected;
- Owner's right to liquidated damages is not relevant as the completion date is not affected provided the contract did not specify a completion date for the particular (delayed) section of the Works with liquidated damages for late completion of the relevant section;
- Contractor may be entitled to financial loss and reimbursement of expenses incurred because of the delay cause by the Owner. Contractor needs to have kept very good records to claim such entitlement.

8.5 Takeaways

Delay is relevant in EPC/Turnkey contracts where it causes a delay in the final completion of a project and thus in the final Taking-Over of the Works by the Owner.

For the Owner the loss is caused by the fact that he can start using the completed works later than planned and therefore loses income and profit.

For the Contractor delay causes loss because he cannot execute the Works as planned and has his personnel and that of his subcontractors standing idle while waiting for the opportunity to start or continue their work.

Delay may be caused by:

- Strikes,
- Failure to grant access,
- Shortage of labour,
- Project Manager's or Owner's instruction,
- Late provision of information,
- Late approval of design documents,
- etc.

Delays may be caused by the Contractor as well as by the Owner and sometimes at the same time; when they occur at the same time it is referred to as concurrent delays.

Where the affected activities are on the critical path and cause the same delay to completion meaning the one without the other, and vice versa, would have caused the delay to completion, they affect the completion date to the same extent. In such case the Owner does not have an entitlement to liquidated damages for delay and the Contractor does not have a right to an extension of time.

Delays may be caused by the Contractor as well as by the Owner sequentially. One delay-causing event may be for the risk of the Owner and another delay may be for the risk of the Contractor. If neither delay affects the time of completion the Owner does not have an entitlement to liquidated damages for delay and the Contractor does not have a right to an extension of time. It is of vital importance to keep very good records.

CHAPTER 9

Testing the works

9.1 Introduction

This chapter describes the phase that follows when the Contractor has finished the Works and wishes to deliver the project to the Owner. At that time, he must show that the Works conform with the Owner's specifications and he is delivering what the Owner requested. This is done by performance testing.

9.2 Performance tests

Performance tests are the tests required to demonstrate the achievement of the guaranteed performance as defined in the contract. The performance tests must be performed and completed before Taking-Over. The Contractor provides the Owner with the Contractor's methodology for conducting the performance test(s) a good number of days, sometimes 90 days before the tests are to begin. The methodology is to be in accordance with Good Practice and the relevant requirements of the contract.

As soon as the Contractor determines that there are no impediments to conducting the performance tests, he must give the Owner notice that he is ready to commence with the performance tests.

The notice specifies the date, location and nature of the performance test to be conducted:

> The Contractor must give the Owner reasonable opportunity to attend all stages of the performance test and such attendance is to be at the Owner's own expense. The Owner's failure to attend any stage of the performance test, through no fault of the Contractor, cannot invalidate the results of the performance test or permit the Owner to allege that the Taking-Over Conditions have not been fulfilled solely on the basis that the performance test which Owner failed to attend was not validly conducted.

Particularly in the chemical industry it is of great importance to have all details taken care of before the testing starts. The composition of the feed stock that is to be used as fuel must be established as compliant with the data given in the Employer's Requirements and on which the Contractor based his design.

Each performance test must be conducted under the authority and direction of the Contractor. If the contract specifies that any stage of the performance test is to be performed by the Employer or a third party appointed by and responsible to the Owner, then the personnel of the Owner or third party are to act under the supervision and instruction of the Contractor.

If the Owner is obliged under the contract or otherwise to fulfil any obligation in order to enable the Contractor to perform a performance test, and the Contractor has notified to the Owner the timing, nature, quality and extent of the Owner's obligation, then the Owner's material failure to comply with its obligation must be treated as a variation. If the Owner fails on more than one occasion to fulfil any obligation in order to enable the Contractor to perform a performance test, the Contractor must give the Owner a final time period of e.g. seven days in which the obligation is to be fulfilled and if the Owner continues to fail to fulfil his obligation within such time then there is cause to consider the performance test as deemed to have been passed and certified, unless, of course, otherwise agreed or otherwise required by the laws of the country where the site is located.

The performance tests are to demonstrate the performance levels of the Works and whether the Works fulfil the minimum performance, if any. In the event that the Works do not achieve the guaranteed performance and the Contractor must execute remedial Works, the Employer must be able to claim a retest to establish that the Works achieve the guaranteed performance.

If the Works fail to achieve the guaranteed performance during the performance test, and/or the retest, then performance-related liquidated damages as set out in the contract are to be paid by the Contractor. The payment of any liquidated damages should not affect the Owner's rights to terminate the contract.

The parties must have agreed and acknowledged in the contract that the amounts of liquidated damages specified in the contract are agreed upon and fixed by the parties because of the difficulty of ascertaining on the contract date the exact amount of damage to the Owner in the event the Works do not conform to the guaranteed performance. The parties then have also agreed that the liquidated damages agreed upon in this contract are applicable regardless of the amount of such damage to the Owner.

9.3 Takeaways

1. The Contractor must notify the Owner well in advance that he is ready to execute the performance tests to enable to Owner to prepare himself to attend the performance testing. If the Owner does not or cannot attend the performance testing through no fault of the Contractor, the Owner cannot claim that the performance tests were not executed validly.
2. Performance testing may take quite some time, so the Contractor must also inform the Owner in a timely manner of the complete schedule of the performance testing. The performance testing is to take place under the authority and direction of the Contractor. Where the contract has

specified that certain parts of the tests are to be executed by Owner's personnel or by a third party appointed by the Owner, such execution must still be under the authority and direction of the Contractor.

3 The performance tests must demonstrate the performance levels of the Works and whether the Works fulfil the minimum performance, if defined under the contract. In the event that the Works do not achieve the guaranteed performance and the Contractor executes remedial Works, the Employer can claim a retest to establish that the Works achieve the guaranteed performance. Such a retest must be subject to the performance test provisions of the contract.

CHAPTER 10

The Taking-Over by the Owner

This chapter describes the Taking-Over procedure.

The Works will be taken over by the Owner on the day that the Taking-Over Conditions have been met or on the date on which the Owner, its agent or designee takes possession and control of the Works.

10.1 Taking-Over Conditions

The Taking-Over Conditions are as follows:

- a The Works have been completed in accordance with the contract, except for minor work which does not affect the safety, operation and operability of the Works and is set out in the Punch List;
- b The Works are fit for the purpose specifically defined in the contract and, if applicable, are capable of being put into operation;
- c The performance tests have been performed and passed in accordance with the Contract, testing equipment has been removed and the plant has been set to the settings specified in the contract or, if no settings are specified, to the settings typically used for operation in accordance with Good Practice; and
- d The consumables, spare parts and special tools required by the contract (if any) are present at the site and are stored in accordance with applicable laws and Good Practice.

10.2 Transfer of responsibility for the Works

On the date of Taking-Over, the Employer takes possession and control of the entire Works or a section of them and is solely responsible for the operation and maintenance thereof and bears the complete risk of loss or damage to the Works or the section that has been taken over. The Take-Over by the Owner does not relieve the Contractor of its obligations during the Defects Liability Period, or from its obligation to perform such other work as remains to be performed after Taking-Over pursuant to this contract and the Contractor must be allowed access to the site and the Works after Taking-Over as needed, to fulfil its remaining obligations. If the Works or any section is to be completed in more than one step or phase, it is advisable to agree with the Owner a reasonable stepped or phased schedule.

When the Contractor believes that he has met the Taking-Over Conditions of the Works or any section in accordance with the contract, he must notify the Owner in the form as set out in the contract, and otherwise any other reasonable form if none is set out.

10.3 Taking-Over procedure

The Owner must, as soon as practicable and in any event within the period agreed in the contract after receipt of the notice of completion, organise to inspect all Works and the results of the performance tests and, within the time period stipulated in the contract, do one of the following:

a Deliver to the Contractor a Taking-Over Certificate for the Works; or
b Deliver to the Contractor a Taking-Over Certificate for a section or sections of the Works; or
c If reasonable cause exists for doing so, notify the Contractor that the Taking-Over Conditions of a section or sections have not been achieved, stating in detail the reasons therefore and the work that it considers still to be done to achieve Taking-Over.

In the event that the Owner determines that the Taking-Over Conditions have not been achieved, the Contractor must promptly take action and perform such work as necessary to subsequently achieve Taking-Over and issue to the Owner another notice of the Works' readiness for Take-Over. If the action or work requires a variation, meaning: if non-achievement of Taking-Over Conditions is the responsibility of the Owner, it must be valued under the variation provision of the contract.

The Owner must issue the Taking-Over Certificate, or the notice that Taking-Over Conditions have not been achieved, as the case may be, within a reasonable time following receipt of the Contractor's notice that the Works are ready for Take-Over. The Taking-Over Certificate shall be retroactive to the Taking-Over Date as defined in the contract. If the Owner fails to issue a Taking-Over Certificate or a notice under the Contract, then the Owner must be deemed to have, and be treated for all purposes as having, issued a Taking-Over Certificate, whether it is actually issued or not as of the Taking-Over Date. The Employer is then not entitled to withhold payments claimed by the Contractor on the basis that the Taking-Over Certificate was not issued or properly issued due to a failure by the Employer.

10.4 The Taking-Over Certificate

Whether it is deemed to be issued or actually issued, the Taking-Over Certificate operates as an acknowledgement by the Owner that the Contractor has fulfilled its obligations under the contract to the extent that the contract calls for such obligations to be performed prior to Taking-Over, except for work (whether minor or not), as the Owner and Contractor agree, remains outstanding, under a Punch List, snag list or other check list of outstanding work to correct defects.

If the Taking-Over Certificate was deemed to have been issued by the Owner, then the Owner is entitled to require the Contractor to still fulfil any Contractor obligations left unfulfilled at Taking-Over. Such work is to be performed at the Contractor's expense.

10.5 Assessment procedures after Taking-Over

To the extent the Owner and Contractor have agreed on certain assessment procedures to be conducted after Taking-Over, such provisions agreed between the parties with respect to such procedures apply. Where no agreement has been made on how the assessment procedures after Taking-Over are to be conducted, they must be conducted in accordance with Good Practice.

If there are no such procedures agreed, then the Contractor must still have access to the site as needed to do necessary work.

If there are reasons, which are not attributable to the Contractor, why any tests after Taking-Over cannot be completed during the Defect Correction Period (or any other period agreed upon by both parties) then the Works are to be deemed to have passed the tests after Taking-Over.

10.6 Relevance of Taking-Over

Upon Taking-Over, the Works are "provisionally accepted" which means that the Contractor has completed the Works and the risk transfers to the Owner, though often the parties agree in the contract on a Defects Liability Period of a certain period. Provisional acceptance becomes Final Acceptance after the Defects Liability Period has passed and any defects, attributable to the Contractor, that have occurred have been repaired. Then Final Acceptance occurs and the Contractor no longer has any liabilities regarding defects in the Works.

10.7 Takeaways

1. The Works are taken over by the Owner on the day that the Taking-Over Conditions have been met or on the date on which the Owner, its agent or designee takes possession and control of the Works.
2. The Taking-Over Conditions should be as follows:

    a. The Works have been completed in accordance with the contract, except for minor work which does not affect the safety, operation and operability of the Works and is set out in a so-called Punch List;
    b. The Works are fit for the purpose specifically defined in the contract and are capable of being put into operation;

c The performance tests have been performed and passed in accordance with the contract, all testing equipment has been removed and the Works have been set to the settings specified in the contract or, if no settings are specified, to the settings typically used for operation in accordance with Good Practice; and

d The consumables, spare parts and special tools required by the contract (if any) are present at the site and are stored in accordancc with applicable laws and Good Practice.

3 On the date of Taking-Over, the Employer takes possession and control of the entire Works or a section of them and will be solely responsible for the operation and maintenance thereof and bear the complete risk of loss or damage to the Works or the section that has been taken over.

4 The Take-Over by the Owner does not relieve the Contractor of its obligations during the Defects Liability Period, or from its obligation to perform such other work as remains to be performed after Taking-Over pursuant to this contract and the Contractor must be allowed access to the site and the Works after Taking-Over as needed, to fulfil its remaining obligations.

5 In the event that the Owner determines that the Taking-Over Conditions have not been achieved, the Contractor must promptly take action and perform such work as necessary to subsequently achieve Taking-Over and issue to the Owner another notice of the Works' readiness for Take-Over.

6 The Owner must issue the Taking-Over Certificate, or the notice that Taking-Over Conditions have not been achieved, as the case may be, within a reasonable time following receipt of the Contractor's notice that the Works are ready for Take-Over.

7 The Taking-Over Certificate shall be retroactive to the Taking-Over Date as defined in the contract. If the Owner fails to issue a Taking-Over Certificate or a notice that Taking-Over Conditions were not met, then the Owner must be deemed to have, and be treated for all purposes as having, issued a Taking-Over Certificate, whether it is actually issued or not as of the Taking-Over Date.

8 Whether it is deemed to be issued or actually issued, the Taking-Over Certificate operates as an acknowledgement by the Owner that the Contractor has fulfilled its obligations under the contract to the extent that the contract calls for such obligations to be performed prior to Taking-Over, except for work (whether minor or not), as the Owner and Contractor agree, remains outstanding, under a Punch List, snag list or other check list of outstanding work to correct defects.

9 Upon Taking-Over, the Works are "provisionally accepted" which means that the Contractor has completed the Works and the risk transfers to

the Owner, though the Contractor remains liable for any defect attributable to him during the Defects Liability Period of a certain period if so agreed in the contract. After expiry of the Defects Liability Period without any defects occurring, the Contractor no longer has liabilities regarding defects.

CHAPTER 11

The Defects Liability Period

This chapter describes what must happen when during an agreed period after Taking-Over certain defects in the Works occur and who will be liable for such defects.

11.1 The Defects Liability Period

The Defects Liability Period is the period after the Taking-Over Date during which the Contractor remains liable for defects in the Works. During this period, the Owner is to notify the Contractor of a defect in design, materials or workmanship of the Works which defect in the Owner's opinion is attributable to the Contractor and during which time the Contractor is obliged to correct such defects at no additional cost for the Owner. This period is often one year in length. If there is sectional Taking-Over agreed, the Defects Liability Period of one year for such section starts with the date of the sectional Taking-Over. If individual items of the plant are repaired or completed during the Defects Liability Period, a new Defects Liability Period of the same duration as the initial Defects Liability Period for those items starts on the date of repair or completion of the Works. It is advisable to limit the Defects Liability Period for any part of the Works to two years after the date of the last Taking-Over Date. This prevents the so-called evergreen defects liability of the Contractor, which would have price consequences because of the ever-extending liability.

The Owner is to notify the Contractor of any defects during the Defects Liability Period as soon as reasonably possible upon such defects becoming apparent. Similarly, the Contractor must as soon as reasonably possible notify the Owner upon becoming aware of defects. During the Defects Liability Period the Contractor is obliged to maintain such goods, materials, equipment and manpower on the site only as is necessary to correct known defects.

The Contractor is not obliged to correct design errors, material choices or workmanship defects attributable to third parties (other than its subcontractors or suppliers) or the Owner, nor is the Contractor obliged, without a variation, to vary or improve Works that, in fact, comply with the contract.

The Owner must provide the Contractor every opportunity and access to complete any outstanding work and correct any defects specified in the Punch List or any defects in the Works not so specified but which have been notified by the Owner to the Contractor during the Defects Liability Period.

11.2 Wear and tear

Normal wear or tear or damage to the Works caused by the Employer's occupation and/or use of the Works or part thereof and/or non-compliance by the Owner or third parties (Operator) with the operation or maintenance manuals supplied (if any) cannot be the responsibility of the Contractor. The Contractor cannot be obliged to repair such wear, tear or damage. It is therefore of great importance that the Contractor will have access to the operational data of the Works kept by the Owner and if such data are not kept or made available, the defects must be deemed to have been caused by the Owner or under Owner's responsibility.

The parties must consult with each other and are to agree the time or times when the Punch List items and/or defects are to be completed and/or remedied. The parties must consult on the best method for remedying any defect, which may include, at the Contractor's option, either: repairing the defect, replacing defective items or reperforming defective work.

11.3 Software issues

If there is a software defect, the remedy may be a software patch, update, upgrade or new software version. If the cost of remedying the defect is disproportionate to the nature of the defect or its magnitude, the Contractor should have the option of offering a reduction of the contract price. The Owner however should not be obliged to accept such offer.

The Contractor should not be liable for defects in software:

1 In case of insignificant deviation of the software from the agreed characteristics;
2 If the defect cannot be reproduced by the Owner in the presence of the Contractor; or
3 If the defect is in open source software.

11.4 Joint inspection

The Owner and the Contractor must carry out a joint inspection of the Works at an agreed time prior to the expiry of the relevant Defects Liability Period, to confirm the condition of the Works. The Owner must immediately notify the Contractor of any outstanding defects to be made good or work required to be executed by the Contractor, all in accordance with the contract.

11.5 Searches for defects

The Owner may need to instruct the Contractor to search for defects either before or during the Defect Correction Period for the section of the Works being considered, When the Owner does, the Contractor must follow such instruction and search for defects. If a defect is found during a search and the defect is the responsibility of the Contractor under the contract, the cost of the work to remedy the defect shall be borne by the Contractor, including the cost of any searches, tests or trials necessary to determine the existence or extent of the defect. In all other cases, the cost plus profit of such work, including any searches, tests or trials, shall be ascertained and paid to the Contractor as if it were additional work and instructed as a variation.

11.6 Reduction of the contract price

In lieu of remedial work, the Contractor must be allowed to offer a reduction of the contract price. The Employer, however, cannot be obliged to accept such an offer.

11.7 Failure to remedy defects

Where the Contractor unreasonably fails, neglects or refuses to remedy defects for which he is responsible, the Owner may upon giving the Contractor the contractual number of days' notice of his intent to have the remedial work done by another Contractor, engage another Contractor to remedy such defects. Subject to the limitations and exclusions of liability in the contract, the Contractor is responsible for the reasonable cost of such remedial work.

11.8 Final Acceptance

Final Acceptance occurs when:

a The Works and Punch List items (if any) have been completed in accordance with the contract;
b The Defect Correction Period has expired and the Contractor's obligation to remedy any defect has been carried out and completed;
c The final as-built drawings and manuals (if any) have been delivered to the Owner; and
d The conditions for Final Acceptance, as agreed between the parties in the Special Conditions of the contract are met.

The Contractor is entitled to receive a Certificate of Final Acceptance when conditions (a) through (d) are met.

Please note that as-built drawings and manuals do reflect any changes made during the Defects Liability Period and are handed over by the Contractor to the Owner after expiry of the Defects Liability Period. Then they reflect the exact configuration of the Works which will come in very handy when years later the Works will have to be revamped and will save a lot of time and money in the preparation and execution of such a revamp.

11.9 Takeaways

1. During the Defects Liability Period, the Owner is to notify, as soon as reasonably possible, the Contractor of a defect in design, materials or workmanship of the Works which defect in the Owner's opinion is attributable to the Contractor and during which time the Contractor is obliged to correct such defects at no additional cost for the Owner.
2. The Owner must provide the Contractor every opportunity and access to correct any defects which have been notified by the Owner to the Contractor during the Defects Liability Period.
3. Normal wear or tear or damage to the Works caused by the Employer's occupation and/or use of the Works or part thereof and/or non-compliance by the Owner or third parties (Operator) with the operation or maintenance manuals supplied (if any) cannot be the responsibility of the Contractor. The Contractor cannot be obliged to repair such wear, tear or damage.
4. It is of great importance that the Contractor will have access to the operational data of the Works kept by the Owner and if such data are not kept or made available, the defects must be deemed to have been caused by the Owner or under Owner's responsibility.
5. The parties must consult on the best method for remedying any defect, which may include, at the Contractor's option: repairing the defect, replacing defective items or reperforming defective work.
6. The Owner and the Contractor must carry out a joint inspection of the Works at an agreed time prior to the expiry of the relevant Defects Liability Period, to confirm the condition of the Works. The Owner must immediately notify the Contractor of any outstanding defects to be made good or work required to be executed by the Contractor, all in accordance with the contract.
7. In lieu of remedial work, the Contractor must be allowed to offer a reduction of the contract price. The Employer, however, should not be obliged to accept such an offer.
8. Final Acceptance occurs when:

   a. The Works and Punch List items, (if any) have been completed in accordance with the contract;
   b. The Defect Correction Period has expired and the Contractor's obligation to remedy any defect has been carried out and completed;
   c. The final as-built drawings and manuals (if any) have been delivered to the Owner; and
   d. The conditions for Final Acceptance, as agreed between the parties in the Special Conditions of the contract are met.

The Contractor is entitled to receive a Certificate of Final Acceptance when conditions (a) through (d) are met.

CHAPTER 12

Contract management

Engineering Contractors take on the following obligations when entering into EPC/Turnkey contracts with their Owners, whether in the oil and gas industry or in the Heavy Civil Industry:

- The obligation to design;
- The obligation to manufacture;
- The obligation to supply;
- The obligation to erect or supervise erection;
- The obligation to construct;
- The obligation to commission;
- The obligation to prove performance;
- The obligation to complete on time;
- The obligation to warrant or guarantee materials, construction and performance;
- The obligation to fulfil all these obligations for the agreed contract price.

12.1 Purpose of contract management

The basic purpose of contract management is to ensure that all obligations are fulfilled and that all rights are enforced. As every project differs from all previous projects, or future projects for that matter, ensuring that all obligations are met and all rights are enforced is mostly a daunting task. Though he is not the project team's ultimate leader, the Contract Manager plays a very important and significant role. It is of the greatest importance that the designated Contract Manager will be involved very early in the process leading to a project, so already in the bidding phase the Contract Manager must be actively involved in scrutinising not only the contract conditions but also the Employer's Requirements and accompanying information. He must work side by side with the engineering department, the procurement department and the execution department, the sales department, the finance department and the company's management so that he has a clear view of all the facets of the project and the company's requirements and must be involved in the decision-making process where management decides whether or not they are going to bid for the project.

Under an EPC/Turnkey contract a Contractor doesn't simply carry out engineering work but in fact a Contractor sells projects and the obligations of the Contractor are as described earlier.

As every project is different, but not unique, one can always learn from previous projects. There may be similarities to previous projects; in each project the Owner may be different, the physical environment of the site, the contract conditions, construction Joint Venture or Consortium partners, subcontractors, etc., all or most may be different, yet the earlier-mentioned obligations must all be met to the satisfaction of the Owner. In that context the contract conditions are extremely important as they describe the obligations of both sides and thus are the evidence of the agreement between the Owner and the Contractor, together with further additional written documents signed by both the Contractor and Owner during the course of the project. As the basic purpose of contract management is to ensure that all obligations are fulfilled and that all rights are enforced, it is imperative that a Contract Manager is appointed to keep all necessary records during the course of the project.

The project organisation charts must include a contract management position, responsible for this function. For small or simple projects, this task may be assigned to the Project Manager.

The effective management and administration of a contract is a time-consuming job requiring day-to-day attention to numerous commercial and contractual details. Therefore, and for large or complex projects, a Contract Manager should be assigned for this position on a full-time basis.

While the Project Manager remains ultimately responsible for the satisfactory performance of the contract, he may, under his control, delegate part of his scope of responsibility to the project's Contract Manager.

The project's Contract Manager must be responsible to administrate:

- The prime contract between the Contractor and the Owner.
- The agreement(s) between the Contractor and his JV or Consortium partners His responsibility extends to assure that:
    - Contractor's contractual obligations are fulfilled.
    - Contractor's rights are enforced.
    - Contractual risks are anticipated and proactive mitigating actions are proposed to the Project Manager.
    - Opportunities in the interest of the Contractor are anticipated and exploited.

The project's Contract Manager also must have the responsibility to approve all contractual documents pertaining to other parties (vendors, subcontractors and licensors) and to check their consistency with the Contractor's contractual obligations towards the Owner.

The project's Contract Manager must have access to the Contractor's legal team to get guidance and advise if and when necessary.

12.2 Contract management program

The Contractor should implement the following program during the execution of the project:

12.2.1 Thorough contract review

To ensure that the contract's requirements and associated risks have been fully assimilated and understood by the project team, a contract review meeting must be held not later than four weeks after the contract has become effective.

The meeting is to be conducted by the Contract Manager and chaired by the Project Manager.

The main issues addressed during the review are the following:

1. The contract base line;
2. Limits of scope of work/services, identification of interfaces;
3. Performance guarantees;
4. Schedule guarantee;
5. Liabilities;
6. Unusual Contractual Provisions;
7. Quality of the Technical Dossier/Endorsement of the Basic Package;
8. Relationships with third parties: licensors, other Contractors, Marine Warranty Surveyor, local authorities;
9. Relationships with partners, if any;
10. Change order procedure.

The Project Manager is to be responsible for further follow-up of actions (if any), as decided during the review.

12.2.2 Preparation of plans and procedures

For large or complex projects, it is advisable to prepare the following documents at the start of the project:

- Contract management responsibilities matrix,
- Contract-required action list,
- Contract-required notices list,
- Flow down clause plan,
- Liability confinement plan,
- A project change order procedure.

12.2.3 Control of the project documentation

Timely and effective written communication is essential to successful management of contract. In case of litigation or arbitration, a successful outcome highly depends on the ability to retrieve information and to produce the right and reliable supporting documents. If prepared contemporaneously of the event subject of the dispute, it will be an invaluable proof to establish the Contractor's entitlement to its request.

12.2.4 Control of contract changes

The change clause usually spells out in detail the procedural requirements to be strictly followed and adhered to by both the Owner and Contractor. Failure to comply may forfeit the Contractor's rights for recovery of money and an extension of time.

The Project Manager and the project team must adhere to the following principles:

- Changes shall be handled with the Owner as they occur and resolved to the satisfaction of both parties quickly.
- When a change arises, put in place a system to record and segregate the man-hours, cost and expenses incurred by the Contractor in relation with this change. The availability of this information is particularly important for an unresolved but disputed change that will ultimately become a claim. This needs to be implemented in very close cooperation with the cost control manager.
- The Owner must be notified of all and any constructive changes. These changes are often conveyed wrongly through, in particular, comments on the Contractor's submittals and they may have large and damaging impact resulting from the "knock-on" or "ripple" effects.

12.2.5 Control of the project schedule

Any prime contract requires that the Contractor prepares and submits a project schedule in some form or another. The purpose of the project schedule is to show the manner and sequences in which the Contractor is planning to carry out the project scope of Works.

It is vital:

1. To be aware of the importance of the Construction Program Manual (CPM) project schedule and to understand that the CPM project schedule is a powerful instrument to demonstrate or defeat the Contractor's rights for requesting an extension of time (EOT) in case of delays and supporting a delay and disruption claim.
2. To develop the CPM so that the Owner obligations are on the critical path because only Owner caused delays are "compensable".
3. To issue timely notification for requesting extension of time (EOT) in case of "excusable" or "compensable" delays.

12.2.6 Anticipation of a possible claim

Contractual terms and conditions in an EPC/Turnkey contract can be very demanding, particularly in a project finance set-up. Owners implement the contract terms and conditions as agreed and written, particularly when in their interests. Owners often have tight budget constraints and under a project finance set-up it is virtually impossible to get Owners and lenders to agree to an increase in costs and an extension of time for completion. The Contractor therefore must prove his case and demonstrate clearly that the additional cost and time overruns incurred are solely attributable to causes for which the Owner is liable.

This new approach needs to be explained to, and understood by, the project team in the early days as the project starts.

12.2.7 Claims management checklist

- A claim is nothing other than the assertion of a right – what one party considers to be its right under the contract, to additional time or money or both.
- A claim is not a dispute, nor need it be contentious. Asserting a right does not mean that the asserting party is aggressive or looking for a fight.
- Construction claims are very common in today's business environment.
- Claims Management is all part of Project Management, and it carries on throughout the lifetime of a project. One must not wait until the very end of the project before deciding to launch a claim – the other party will also have left the site and will have little or no interest in spending management time on a claim, and by waiting one will have lost leverage.
- Administer the terms of the contract properly.
- Make sure that sufficient notice has been given of the intention to claim within the time limitations set out in the contract.
- The claim must be accurate, realistic and convincing. Above all, the claim must be based on the contract.
- The claim must include a causal connection: an explanation of what the other side has done wrong and what detrimental effect the wrongdoing has had for the claimant.
- Be aware of changes and variations. Be very aware of constructive changes.
- Assess your delay entitlement fully – select and adjust your baseline schedule.
- Always consider float, concurrency and mitigation.
- Distinguish between disruption costs and prolongation costs.
- Keep acceleration costs separate.

One may be on the receiving end of a claim. In that case there is no reason to panic. A Counterclaim is not a defence against a claim! The best response is to deal properly with any claim that one receives: gather the project team and the project documents and go over the particulars of the claim, the causal connection, the view of the project team members directly involved with the particulars of the causal connection and calmly discuss their views and as a Contract Manager be critical in listening to their comments. It is most prudent not to deny a good solid claim from the other party.

It is always best to settle a claim whenever possible, but settling at any price is not sensible and should, of course, be avoided.

It is always better to consider mediation before arbitration, and it is really better to use Dispute Boards whenever possible!

12.2.8 Lessons learnt

Lessons learnt are valuable assets. The Contractor and Owner do themselves a big favour by setting up a system to record lessons learnt from projects and to regularly have training sessions with their staff so that all within their staff can benefit from experiences from the past. It means growing as an organisation and becoming better at what one does.

12.3 When defects occur

The Contractor must perform the work free of defects. What exactly is a defect? Is it only when (a) part of the Works is broken or not functioning? Defective can be more than that: what standard does the contract, the law or both set for the work? Installed equipment may not be compliant with environmental regulations. The Contractor's failure to ensure compliance with local environmental laws may result in a defective work product.

European Civil Law approaches the concept of defect by comparing the as-built condition with the as-required (by contract or by law) condition. If there is a difference between these two criteria, which results in an impact on the average use or the intended use of the equipment at issue that is not only marginal, then it is likely that the subject will be judged defective. It is no defence that certain products are never free from defects such as e.g.: software products.

Most engineering contracts oblige the Contractor to make-good, to remedy a defect that occurs after Take-Over by the Owner. This so-called Defects Liability Period commences when the Owner starts production after he accepted the Works. Often control of the equipment and the passing of the risk of loss or damage passes from the Contractor to the Owner at acceptance by the Owner. When a defect occurs, the Contractor must have access to the operations data on the use of the Works by the Owner after acceptance of the Works. It wouldn't be the first time an Owner uses the Works beyond the specifications given to the

Contractor in the Employer's Requirements, on the basis of which the Contractor designed and built the Works.

The contract may give the Owner the right to inspect any work in progress and to request remedial action at such early stage before any kind of acceptance of the work by the Owner has taken place. Such inspection may have positive effects as the forced review by the Contractor may result in the review and supervision of his subcontractors. The Contractor may spot problems at an early stage when remedying them is usually less costly.

If a breach of warranty is a material or fundamental breach, the non-breaching party may consider terminating the contract all together and claim full damages. This would be the moment to engage the Dispute Adjudication Board to establish that the other party committed a material or fundamental breach. If not, the non-breaching party suddenly will be in breach of contract and may expose itself to substantial damages.

12.3.1 Subcontractor's warranties

Also, the subcontractor will have to provide warranties. The Contractor is liable to the Owner, and usually there will be no contractual relationship between the Owner and the subcontractor from the contract between Owner and Contractor. Thus, if there is a defect in the subcontractor's performance, failure to act by the subcontractor is failure by the Contractor. Therefore, every Contractor must make sure that the contractual provisions and, in particular, the warranties are "back to back" to the contract between the Contractor and the Owner. The subcontract is to be a mirror image of the main contract between Owner and Contractor. Beware, however, of certain problems that such a pure mirror contract may cause.

When the main contract allows the Contractor, for example, five days to investigate and, if necessary, commence remedying an alleged breach of warranty and the subcontract allows the subcontractor the same time period, the Contractor must have a perfect administration that will handle the Owner's notice of a breach of warranty the very same day and also give notice to the subcontractor. If the Contractor fails to do so and gives notice to the subcontractor only two days later, the subcontractor could satisfy its contractual obligation, for example, on the fifth day after receipt of notice while this would establish the seventh day in the relationship between Contractor and Owner. After the fifth day, the Owner then could exercise its right of engaging another Contractor to remedy the defect, at the cost of the Contractor.

It is therefore prudent to have the following provision in the subcontract, where the project is e.g. for the realisation of a cracker furnace in the Oil and Gas Industry but which may be appropriate in other industries as well, where components are manufactured and installed by subcontractors:

> If the Owner does not allow the Contractor an opportunity to carry out the Final Hot Performance Tests of all Facility Components within 12 months after the Preliminary Plant Acceptance, the Final Plant Acceptance Certificate shall be deemed

> to be issued at the expiration of the said 12-month period, unless, during the said 12-month period any Final Hot Performance Test of any Facility Component has been conducted and the Facility Component has failed to meet and satisfy the Hot Performance Test.

Imagine that the Contractor has a subcontractor whose scope of work is limited to one so-called Facility Component. The contract between subcontractor and Contractor is back-to-back to the contract between Owner and Contractor. Furthermore, the Contractor conducted a so-called Final Hot Performance Test within the 12-month period within the contractual relationship to the Owner, and failed. The Final Hot Performance Test was conducted in a Facility Component that is not part of the subcontractor's scope of work. There was no Final Hot Performance Test opportunity for the subcontractor within the 12-month period.

What is now the consequence? Should the Final Plant Acceptance Certificate be issued at the expiration of the said 12-month period with regard to the subcontractor's scope of work?

And what about payment for costs incurred because of unjustified Owner claims for breach of warranty? Must the Owner pay the costs including labour costs plus mark-up (profit) if the Contractor investigation does not reveal any breach? To avoid any uncertainties, the Contractor better include a provision that covers such costs. As a welcome result of such a clause an Owner will likely not flood the Contractor with frivolous warranty claims.

Then, what about any alleged Owner right to perform emergency repairs? Is there such an Owner right to perform emergency repairs? Many contracts do not touch the issue. In practice, however, a Contractor will not stay with the work for the warranty period to provide 24/7 coverage in case of any alleged breach of warranty. If the Owner has the right to perform the right of self-help in cases of so-called emergency repairs several issues will come up:

> Was the performed repair an emergency repair? Contractors should be aware that Owners that are keen to increase output consider everything an emergency repair that negatively affects the productivity.

Even if the Contractor happens to agree to the qualification of "emergency repair", the cause for the alleged breach of warranty may simply be improper operation or inadequate maintenance. Both causes will be hard to prove if the Owner has the right to go ahead and repair the damage before the Contractor had an opportunity to investigate the cause for the failure. In such case the Contractor must have access to the operational data of the Owner.

12.3.2 Variation orders

Almost every contract will have a provision within its terms that allows the Owner to change or vary the quality and/or the quantity of the project Works. The Owner requests a change or variation to the project, and the Contractor carries out that change and gets paid accordingly. Although the principle is simple, changes to the scope of work often are a significant source of conflict within a construction project. Such conflict arises from two main sources:

- Whether or not the work in question is in fact a change to the original contracted scope of work, and
- Determining a price or any other compensation due for the change as executed.

Although most contracts will define what a variation is, and how it is to be administered, there are frequent arguments about how much is to be paid for carrying out the variation, or also about whether the "changed" work is already part of the Contractor's original scope of work. Often Owners will argue that the work is already considered to be part of the original scope of work and that the "change" has to be carried out for no additional cost.

The execution of a variation order under the contract terms is not a claim. It is simply part of the normal administration under the contract terms and the change – and the additional payment and additional time for completion related thereto – is carried out in accordance with the same contract provisions as the rest of the project Works.

A contractual claim may arise if there is a disagreement between the Owner and the Contractor as to whether the "changed" work is already in or outside of the original scope of work under the contract. Indeed, this is very often the main cause of disputes between the parties. The Contractor wishes to be paid and get additional time for carrying out work which he considers is not included as part of his lump sum price and fixed time for completion. The Owner, on the other hand, considers that the work is already included under the contract and that it is to be executed as part of the original contract terms and at no additional price and time for completion. This then leads to a dispute between the parties and sometimes this can be difficult to resolve.

Changes come in different forms. In their simplest form, there are basically three types of change:

- Changes ordered by the Owner;
- Changes made by the Contractor, which are not instructed; and
- Changes required by a third party (e.g. the authorities).

12.3.3 Changes ordered by the Owner

Most forms of contract will contain terms allowing the Owner to order variations to the original scope of work. Such terms will normally determine the extent of the variation the Owner is allowed to order. Typically, such variations include additional work, the omission of work, a substitution or alteration, or changes in the quality of the work. It is advisable to have a provision in the contract that limits the Owner's ability to order variations beyond a given percentage of the contract price and/or beyond a given number of variation orders. Many Owners do not see that variations to the scope of work are disruptive and often have far more consequences for the Contractor than the Owner can foresee; the Owner than sees it as an attempt of the Contractor to bargain for more money to compensate for the Contractor's own inefficiencies.

Many disputes regarding changes ordered by the Owner stem from either an argument over whether the contract allows the Owner to make such a change or over the price to be paid for the work done and the addition to the time for completion of the project. Most forms of contract will state that a Contractor must comply with a written instruction from an Owner, therefore a Contractor will have little choice but to do what he is asked. However, it is Good Practice to demand a written explanation from the Owner describing the contract powers he is relying on in issuing a particular instruction. If a Contractor should disagree with the Owner's opinion when given, a written notice of claim to that effect should be sent to the Owner.

12.3.4 Changes made by the Contractor

Generally, it is unwise for a Contractor to carry out any changes to the scope of work without a written order from the Owner to do so. Sometimes, however, it becomes necessary to carry out a change ordered verbally by the Owner or his authorised representative. Some contracts will allow such a verbal instruction to stand provided that the Contractor confirms the verbal instruction in writing within a certain period of time. Even if the contract does not provide such terms, each time a Contractor believes he has or will carry out changes to the scope of work (not caused by his own fault) a request for a written instruction must be sent to the Owner.

12.3.5 Changes required by a third party

Changes required by a third party, for example, a change required by a local authority, or planning or fire department or similar, may be regarded as a change entitling the Contractor to additional payments. The risk for such changes is to be defined in the contract to the effect that the Contractor has or is deemed to have included such risks in his price for the contracted work if he is aware of the local authorities' regulations and/or requirements, but not of any changes to such regulations or requirements.

Arguments will develop in cases where such risks are not clearly stated. In such cases, a claim for payment may succeed, but only if in all the circumstances the Contractor could not have been expected to have foreseen, and allowed for, the risk in question in his contract price and time for completion.

These, however, are the straightforward and usual types of changes that are to be found in a construction contract. Less obvious changes, however, are also to be found in a project.

12.3.6 Constructive changes

A constructive change is a change not recognised specifically as such by the parties. The essence of a constructive change is an act or a failure by the Owner that, even though not recorded as a formal change to the contract, has the effect of requiring the Contractor to change the way in which the contract is to be carried out. Sometimes constructive changes may be confirmed in writing by the Owner, but when done so, they tend to be described as "for information only" or "at no additional cost".

In effect, what this means is that the Owner:

- Instructs the work to be carried out;
- Refuses to agree to the value of the changed work;
- Does not think that it is a variation or change to the scope of work; and
- Will attempt to value the change at zero.

Constructive changes also occur where the contract requires something to be done by the Owner, but the Owner doesn't do anything at all. In the face of such inaction, the Contractor may find that he is forced to carry out the action that the Owner failed to do, simply in order to progress with the work and not have his own scope of work (and costs) increased. A constructive change may also be implied where additional work is done with the Owner's knowledge and approval, even though no actual instruction to do the work is made by an authorised representative of the Owner.

Since, in these circumstances there is no formal act of issuing a change order by the Owner, the Contractor must recognise and identify the conduct of the Owner that constitutes a constructive change order. If a Contractor does not do this, then he cannot obtain compensation for his increased costs because the burden of proof of the Owner's conduct is on the Contractor.

The common types of constructive changes are to be found in:

- Defective specifications;
- Changes in method of performance;
- Over-inspection;
- Rejection of conforming work; and
- Defective materials or property provided by the Owner.

12.4 Defective specifications

A Contractor is entitled to rely upon method specifications, and if the specification is defective, then the Contractor is entitled to extra compensation for the costs in correcting the specification. The Contractor is not responsible for the defective result if the methods and materials stated do not produce the desired results; however under European Civil Law he will have a legal obligation to warn the Owner of such fact.

12.5 Changes in method of performance

If a reasonable interpretation of the contract specification allows the Contractor to choose between various methods of performing the work, then the Contractor has the right to choose the least expensive method. If, thereafter, the Owner directs that the work be carried out in a different manner, then a constructive change has been ordered, and the Contractor will be entitled to the additional costs incurred in carrying out the work in the manner now specified by the Owner.

The Contractor has the right to choose the method of performance where no particular method of performance is specified in the contract as well as when two or more optional methods are specified. In either case, when the Owner directs the Contractor to use a more expensive method than the least expensive method available, a constructive change has occurred and the Contractor becomes entitled to claim for the additional costs incurred.

There must be, of course, an instruction given by the Owner – if the Contractor merely volunteers to use a more expensive method of performance, the Contractor cannot recover the additional costs. An acceptance of a suggestion by the Owner may not constitute a constructive change.

12.6 Over-inspection

Over-inspection may take the form of changing the scheduling or frequency of inspections or requiring a higher standard of performance as a result of greater inspections.

In order for there to be a constructive change because of over-inspection, the inspector must be authorised to issue change orders or his action must be ratified by the Owner. The Owner may not be bound to compensate the Contractor for a more expensive method of inspection than that stated in the contract unless it is directed by an authorised representative of the Owner.

12.7 Rejection of conforming work

A constructive change can arise when the Owner unjustifiably rejects the Contractor's work that conforms to the original contract, thus requiring the Contractor to carry out some rework.

Some contracts require the Contractor to follow the specifications using "first-class workmanship" which has been interpreted to mean skilful average work conforming to industry standards. The determination as to whether the original work carried out was to a first-class level of workmanship may be influenced by whether or not the Contractor followed the recommendations of the manufacturer.

12.8 Defective materials or property provided by the Owner

Additional work carried out by the Contractor because of defective property, equipment or materials provided by the Owner under the contract also gives rise to a compensable constructive change. Consequently, the Contractor can claim for the additional costs caused by the defective materials or property supplied by the Owner.

An Owner who contractually agrees to furnish property must provide it suitable for the intended use under the contract. If the property or materials supplied by the Owner is not suitable, then the Contractor is entitled to claim for the additional cost incurred in purchasing suitable property or materials, or for the additional costs of making the unsuitable materials suitable for his performance.

12.9 Consequential changes

Consequential changes are in fact just another form of constructive change, but they can be more difficult to identify. Consequential changes represent additional work or costs that occur as a direct consequence of some other more obvious change. The cost and time effects of a consequential change include all those that can occur as a result of any other change i.e., direct costs and/or delay to progress and disruption costs and/or delay to completion and delay costs.

Whilst the Contractor would normally include all of the consequential effects of a change in his notice for a constructive change, it may not be possible or foreseeable until somewhat later when the entire consequences of a change become apparent. Whenever the effects of a consequential change become clear, the Contractor must immediately notify the Owner and follow the procedures for monitoring a constructive change.

12.10 Constructive acceleration

The US Corps of Engineers defines constructive acceleration as:

> An act or failure to act by the Owner without recognition that the Contractor has encountered excusable delay because of such act or failure to act, for which the Contractor is entitled to an extension of time and which required the Contractor to accelerate his program in order to complete the contract requirements by the existing contract completion date. This situation may be brought about by [the Owner's] denial of a valid request for a contract time extension or by [the Owner's] untimely granting of a time extension.

That is a good and workable definition.

The implied instruction to accelerate is probably one of the more difficult situations to deal with within project and claims management. In today's execution of ever more complex and complicated projects, particularly in regard to the provision and erection of specialist equipment, the doctrine of constructive acceleration has become one of the more contentious issues in recent years. There are those who believe that constructive acceleration forces contractual interpretation to focus on the business efficacy of a Contractor rather than rely upon the courts to interpret the dispute along the lines of old fashioned doctrines of classical contract theory.

In so far as a Contractor's work has increased as a result of an imposed change or where delay to the progress of the work has been caused by neutral causes, and where no extension of time has been granted, the Contractor will find himself in a very difficult situation.

Should he accelerate and incur additional costs, and thus avoid the unjust deduction of liquidated damages, or should he continue in the best way that he can, minimising additional costs but suffering the deduction of LDs?

If the Contractor just struggles on and argues about his right to claim costs later on, his right to claim will be measured by whether or not any overrun of time is driven by an excusable delay. If the Contractor accelerates and overcomes the delay, then in order to recover his incurred costs, the Contractor would have to prove not only that there was an excusable delay, but also that without such acceleration he would have been in delay through no fault of his own and that as a result of the acceleration he has suffered a quantifiable loss.

However, most Contractors find that continuing the work in the hope of fighting against wrongfully deducted LDs, with the additional legal costs, has a far more disastrous impact upon their cash flow situation than increasing the resources needed to complete on time if they can.

The circumstances that usually give rise to a claim for constructive acceleration are where the Contractor appears to be in default and the Owner presses the Contractor to complete on time. The Contractor will then incur additional costs as a result of enhanced performance, and it is subsequently found that the Contractor had been entitled to an extension of time in the first place, as his delay was in fact caused by an act or failure to act by the Owner.

The doctrine of constructive acceleration has been well recognised by American and Canadian courts, though they are less common in Europe. The following six elements are those that the US courts use to identify whether constructive acceleration has occurred:

- There must be an excusable delay;
- Notice must have been given of the delay, and if required by the contract, an extension of time must have been requested;
- The extension of time must either have been refused or not given within a reasonable time;
- The Owner (or Main Contractor in a Contractor/subcontractor relationship) must have acted by coercion, direction or in some other manner that can be construed as an instruction to complete within the agreed contract period;
- The Contractor must actually accelerate his performance; and
- The Contractor must actually incur a loss as a result of the acceleration.

If these six components are in place, then the Contractor will have a reasonable chance in claiming for recovery of his additional costs due to constructive acceleration.

One would expect that if the Contractor can show these six elements of constructive acceleration to a Dispute Adjudication Board or an Arbitral Tribunal under European Civil Law the result would be the same.

12.11 Valuation of changes

Having agreed that a change – constructive or otherwise – is appropriate, it is necessary to evaluate or price the cost of the changed work, be it an increase or a decrease to the contract price. It is surprising how many major international engineering projects contain very little guidance on how to price changes. Many such contracts have provisions stating that a price is to be submitted by the Contractor to the Owner before the work is executed but contain no definite provisions on pricing such work.

The price to be agreed for any changed work must be fair and reasonable. What is fair and reasonable will depend on the circumstances. Essentially, any changed work which is of the same character and is executed in similar conditions to that envisaged by the original contract is to be priced at contract prices or prices similar to contract prices. If the changed work is of a different character and/or is executed in different conditions, new prices that fairly reflect such changes must be agreed.

It is reaching agreement over the cost of changed work for which the contract's list of unit prices does not apply that causes most difficulties and disputes over the pricing of change orders. There are a number of matters that need to be considered:

- The type (character) of additional or different work:
  - An Owner may order additional work for which no price ever existed in the contract. A new price entirely needs to be agreed.
- When the work was carried out:
  - Extra work may have been demanded by an Owner to be carried out at a time not convenient to the Contractor. Disruption to the normal progress of the Contractor's work may have occurred, for example, by the removal of certain workers from their normal routine to carry out the extra work. This may result in a delay to the normal work because of the work stoppage and the delay in restarting the normal work. The cost of such disruption would therefore have to be included in the price for the extra work or a loss would result.
- The circumstances under which the changed work was carried out:
  - An Owner may demand that a particular piece of unexpected extra work be carried out in a winter period on site. Because of adverse weather conditions, the work may take longer than normal and hence cost more. This needs to be reflected in the price agreed for the extra work.
- The number of changes ordered:

- Studies have shown that if a Contractor has to carry out a great deal of changes, the productivity of his work force will fall. This needs to be taken into account when pricing a change.

In addition to the earlier points, care is needed in managing time and unit rates that may have been included in a contract for pricing changes. For example, if a subcontract has such provisions it should be remembered that these unit rates might be considerably higher than the rates used in pricing the contract work. Such time and unit rates are normally used for small portions of work, which under normal circumstances would be unprofitable. If a change order contains substantial work, time and material rates would be inappropriate and a new rate must be agreed.

Care also needs to be taken over the build-up of prices for additional work. A price contains different elements: labour, plant, materials, profit and overhead. A certain proportion of the overhead cost will include a fixed element. Those fixed cost elements need to be identified and removed from the overhead allowance priced for in the price of the additional work or recovery will be made twice.

This is true for time also!

In addition to agreeing the price to be paid or deducted for a change, consideration also needs to be given to the time for performance or completion. It is sometimes astonishing how often significant changes are made to a contract scope of work without any reference to the need for additional time. Carrying out extra work may very well delay completion of the project. Most forms of contract allow a Contractor additional time for delays caused by executing additional work. Some may even demand that an earlier completion date be set if the change ordered results in the omission of work.

When agreeing a price for a change order, any time effects must be agreed at the same time. In addition, any prolongation costs incurred or likely to be incurred as a result of the delay caused must also be agreed as part of the settlement of the change order price.

From the earlier paragraphs it is obvious that contract management covers all of:

- Evaluation of the contract conditions included in the invitation to bid,
- The preparation of the bid,
- The negotiations with the Owner regarding the bid,
- The contract execution,
- The testing and hand-over.

12.12 Claims

The definition of a claim is:

> a request, demand, application for payment or notification of presumed entitlement to which the Contractor or the Owner as the case may be, rightly or wrongly at that stage, considers himself to be entitled and in respect of which agreement has not yet been reached.

How claims are presented and received is fundamental to their success in being found permissible under the terms of the contract under consideration. This is what is called "entitlement under the contract", and this means that if a Contractor believes that he is entitled to be paid extra money and/or be given more time to complete, then he must show that his right to present a claim stems from the contract terms, or under the relevant law.

A Contractor is not entitled to claim just because he suffers a loss or because his budget has been exceeded. Only if he can justify his entitlement under the contract terms will the other party take serious notice.

In order to have a successful claim the Contractor must first establish that he is fully entitled to recover the sums or time that he intends to claim. The claim must be presented in proper contractual terms and it must show that under the contract, the other party is responsible for the delay or additional costs involved.

Without entitlement under the contract, there can be no claim, unless the applicable law creates an entitlement.

A frequent area of confusion is with regard to an extension of time (EOT) for completion. Some Contractors consider that if there is a delay during the progress of the Works, this then translates into a delay in the completion of the project – this confusion can also exist regarding:

- A disruption to the execution of the Works, and
- A delay in the completion of the Works.

Generally, it is only a delay to the completion of the Works that allows for an EOT, though it must be said that an event, or action by the Owner, leading to a delay (and the reasons for the delay explained) must be notified by the Contractor to the Owner forthwith upon discovery of the action or event.

Claims arise when one of the parties to a contract fails to comply with one or more of its obligations. That failure may be a failure to do something that the contract states is his responsibility or is a legal requirement. Or it may be a failure to prevent something from happening. When a party to a contract fails to comply with an obligation it is known as a breach of contract. There are two principal types of breach of contract:

- A technical breach of contract: this is a breach of contract that does not fundamentally affect the functioning and operation of the contract.
- A fundamental breach of contract: this is a breach that is so severe that it goes to the heart of the contract, thus making it unworkable.

12.13 The causal connection

Another very important ingredient that is required for an entitlement to claim to be successful is the causal connection. The causal connection links the event that created the subject of the claim with the party responsible for that event. For example, if an Owner fails to provide materials and services at the time required by the contract, then the causal connection links the Owner's responsibility for providing these materials and services to its failure (or breach) to provide them.

The connection between the Owner's obligation to provide the materials and the subsequent unavailability of the materials when required by the contractual time line will then entitle the Contractor (provided he notified in a timely manner the Owner of the Owner's dereliction of duty and that dereliction directly causing the Contractor delay, the causal connection) to claim for the additional time whilst waiting for the Owner to provide the materials.

12.14 Notices

The submission of timely and adequate notice is fundamental to a successful claim. Sometimes Contractors either do not want to create an adversarial working environment but as the French say: *C'est le ton qui fait la musique* ["It is the tone that decides the music" or "It is not what you say, it is how you say it"]. It is often quite arguable that by sending a notice to the other party, you are doing them a favour; you are drawing their attention to something that is going wrong under their responsibility, and you are giving them the opportunity to correct it before the negative effects increase.

Notice periods are usually specified within the contract, most frequently in respect of change orders for additional work, or in specific regard to claims themselves. Some contracts make it very clear that, for example,

> If and when it becomes apparent that the progress of the Works is being, or is likely to be, delayed the Contractor shall forthwith give written notice.

The reason for giving such notice by the Contractor in such a situation is primarily to get the event or act corrected by the Owner thus allowing the Owner to resolve or mitigate the effects of the problem, and if the Owner fails to correct, to enable the Contractor to assert his right to claim in a timely manner during the life of a project.

The notice need not be complicated. Although the specific requirements of the notice may be set out in the contract, there is usually a minimum amount of information to be inserted in the notice:

- The relevant event,
- The date the event took place,
- Its cause,
- Its likely duration, and
- Its potential impact.

It is not always necessary for the notice to identify the period by which the duration of the contract period is likely to be extended. Indeed, at the time of giving notice, it is possible (sometimes very likely) that the Contractor may have no firm idea as to the extent of delay suffered by the contract progress or the additional cost involved in overcoming the effects of the act or event. This will only be subsequently ascertained once the claim itself, with all its ingredients, has been fully researched and compiled.

12.15 Records

Keeping good records is of vital importance; it is one of the primary tasks of the Contract Manager to make sure that good records are kept by the project staff.

Clear factual evidence is crucial to a successful claim. Good records help to avoid confusion, and they help in reaching settlements by defining facts, events, roles and responsibilities. They also help to co-ordinate activities and also help to reduce the incidence of disputed claims.

A study carried out in the UK recently identified unclear or missing information as being a serious cause of construction defects and also as a recurring cause of disputes themselves. Predictably, poor Project Management has often been the cause of poor record keeping, but experience in construction disputes demonstrates that the earlier facts are established and recorded, the earlier any disruptive situation can be corrected by the party responsible and if not, a claim is likely to be settled in the face of the evidence in good records, which in turn leads to lower litigation/arbitration costs and more successful settlements.

In general, a justifiable cause and effect analysis of entitlement to compensation for additional time or costs is almost impossible to achieve without adequate, contemporaneous project records. In the presentation of a claim, it has been said[1] that:

> A party to a dispute, particularly if there is arbitration, will learn three lessons (often too late):
>
> 1 The importance of records,
> 2 The importance of records, and
> 3 The importance of records.

The nature of the records will vary from project to project and from claim to claim; an analysis of the following items will normally be required:

- The original tender,
- Measured or specific areas of work,
- Program and time schedules,
- Cost records,
- Correspondence,
- Records of site meetings,
- Site diaries,
- Photographs,

1 Max William Abrahamson: *Engineering Law and the I.C.E. Contracts* (4th Edition, E & FN Spon, London, 1985).

- Specifications,
- Change orders, and
- Drawings.

In a complex claim, project documentation becomes the primary source material used to develop the relationship between the causes and effect of delay. Therefore, the success of presenting any form of claim for delay is very dependent upon the availability of adequate records.

12.16 How to preserve one's entitlement

Preserving one's entitlement to a claim is a management function. One must realize that not having claimed full entitlement until it becomes apparent that losses are being incurred is far too late. If a problem is not noticed until it affects cash flow and income it may be very difficult to recover at a later stage. When one leaves the "chess game" behind and focusses on the achievement of the project's objectives and the other party, no matter if it is the Owner or the Contractor, receives timely warning of something going wrong that is under their responsibility and it is corrected the risk is mitigated and the project objectives are no longer endangered.

A management early warning system should be introduced at the start of the project. This does not have to be expensive or complicated and can simply consist of:

- Continuous review of minutes of meetings including internal meetings. Examine, for example, minutes from the engineering, procurement, QA/QC and subcontract departments. Define apparent problems, and then define why they exist.
- Continuous review of all site instructions and change orders. Ensure that the planning department properly advises on the impact of each and every change order on top of the impact of earlier change orders. Define apparent problems, and then define why they have been caused.
- Continuous reviews of all correspondence to look out for problems at as early a stage as possible. Define apparent problems, and then define why they are a problem.
- Regular feedback from planners and schedulers, indicating deviations between short term program objectives and master schedules including overall milestones. This requires continuous review of the original schedule activity durations that were based upon "project standard" objectives and durations as compared with objectives and durations that are actually experienced. Define apparent problems and then define why they are a problem.
- Incorporate into the normal reporting procedures for every department a separate section for "change orders, extra works, claims and any other problems" and insist that each line manager and Project Manager comments

under this heading, even if he/she has no problems they are aware of. If the manager decides no claims exist at present, he/she should say so.

12.17 Be sure to claim entitlement in a timely manner

Do not leave claims until the end of the project. This counts for both the Owner and the Contractor. If one does, it means loss of leverage and there will be no incentive for the other party – Owner, Contractor or subcontractor – to reach a reasonable settlement with the claiming party. In addition, allowing a number of small claims to accumulate during the life of the project will add to the ongoing every-day difficulties and cause a major breakdown in communications. More importantly, if you delay the submission of your claim entitlements until the very end of the project, your cash flow and income become seriously distorted and this may have serious detrimental effects on your balance sheet.

Do not be misled into thinking that a claim will upset or destroy a good working relationship with a party if a proper claim entitlement is presented. If, by any chance, such a working relationship is adversely affected by the presentation of a proper claim entitlement, then that relationship was always doomed to fail in any event. More usually, a party may very well have a higher regard and respect for a party who presents clear, justifiable arguments that are strictly in accordance with the contract. Furthermore, an honest discussion between the parties makes for a better working relationship. If insurmountable differences do occur at site level, then these matters must be deferred to a higher level of management – or to external consultants – and this enables the site level staff on both sides to work in harmony without the friction of ill-feeling and quarrelsome arguments.

12.18 Takeaways

1. Engineering Contractors take on the following obligations when entering into EPC/Turnkey contracts with their Owners, whether in the oil and gas industry or in the Heavy Civil Industry:

   - The obligation to design;
   - The obligation to manufacture;
   - The obligation to supply;
   - The obligation to erect or supervise erection;
   - The obligation to construct;
   - The obligation to commission;
   - The obligation to prove performance;
   - The obligation to complete on time;
   - The obligation to warrant or guarantee materials, construction and performance;
   - The obligation to fulfil all these obligations for the agreed contract price.

2 The basic purpose of contract management is to ensure that all obligations are fulfilled and that all rights are enforced.

3 The Contract Manager is responsible to administrate:

- The prime contract between the Contractor and the Owner;
- The agreement(s) between the Contractor and his JV or Consortium partners. His responsibility extends to assure that:
    - Contractor's contractual obligations are fulfilled;
    - Contractor's rights are enforced;
    - Contractual risks are anticipated, and proactive mitigating actions are proposed to the Project Manager;
    - Opportunities in the interest of the Contractor are anticipated and exploited.

The project's Contract Manager also must have the responsibility to approve all contractual documents pertaining to other parties (vendors, subcontractors and licensors) and to check their consistency with Contractor's contractual obligations towards the Owner.

4 The contract management program comprises:

1 Thorough contract review,
2 Preparation of plans and procedures,
3 Control of the project documentation,
4 Control of contract changes,
5 Control of the project schedule,
6 Anticipation of a possible claim,
7 Claims management checklist,
8 Recording lessons learnt.

5 When a defect occurs, the Contractor must have access to the operations data on the use of the Works by the Owner after acceptance of the Works. It wouldn't be the first time an Owner uses the Works beyond the specifications given to the Contractor in the Employer's Requirements, on the basis of which the Contractor designed and built the Works.

6 A contractual claim may arise if there is a disagreement between the Owner and the Contractor as to whether the "changed" work is already in or outside of the original scope of work under the contract. Indeed, this is very often the main cause of disputes between the parties – the Contractor wishes to be paid and get additional time for carrying out work which he considers is not included as part of his lump sum price and fixed time for completion. The Owner, on the other hand, considers that the work is already included under the contract and that it is to be executed as part of the original contract terms and at no additional price and time for completion.

7 Changes come in different forms. In their simplest form, there are basically three types of change:

- Changes ordered by the Owner;
- Changes made by the Contractor, that are not instructed; and
- Changes required by a third party (e.g. the authorities).

8 The definition of a claim is:

> a request, demand, application for payment or notification of presumed entitlement to which the Contractor or the Owner as the case may be, rightly or wrongly at that stage, considers himself to be entitled and in respect of which agreement has not yet been reached.

How claims are presented and received is fundamental to their success in being found permissible under the terms of the contract under consideration. This is what is called "entitlement under the contract", and this means that if a Contractor believes that he is entitled to be paid extra money and/or be given more time to complete, then he must show that his right to present a claim stems from the contract terms, or under the relevant law. Without entitlement under the contract there can be no claim, unless the applicable law creates an entitlement.

9 Another very important ingredient that is required for an entitlement to claim to be successful is the causal connection. The causal connection links the event that created the subject of the claim with the party responsible for that event.

10 The submission of timely and adequate notice is fundamental to a successful claim. It must contain:

- The relevant event,
- The date the event took place,
- Its cause,
- Its likely duration, and
- Its potential impact.

11 Keeping good records is of vital importance; it is one of the primary tasks of the Contract Manager to make sure that good records are kept by the project staff.

Clear factual evidence is crucial to a successful claim. Good records help to avoid confusion, and they help in reaching settlements by defining facts, events, roles and responsibilities.

12 Do not leave claims until the end of the project. This counts for both the Owner and the Contractor. If one does, it means loss of leverage and there will be no incentive for the other party – Owner, Contractor or subcontractor – to reach a reasonable settlement with the claiming party.

CHAPTER 13

Working in a Consortium or a Joint Venture on a project and the difference between the two

13.1 The reasons for forming a Joint Venture or a Consortium

Parties that want something built or those who are prepared to build that something often form a Joint Venture or a Consortium, as the case may be, for various reasons. The main reasons may be on the Owner's side to share the risks of the project with other parties that have an interest in the project and the project's yield, thereby reducing one's own total risk exposure with regard to the project through the pooling of investment capital. On the Contractor's side the reason can also be risk sharing and thereby reducing one's own risk exposure regarding the project. Additionally, working in a Joint Venture or a Consortium allows for pooling of resources; the project may be too big for each individual Contractor, but together they will be able to build it by sharing the risk and pooling personnel and equipment. They form a Joint Venture when they are in the same business but a Consortium if they are rooted in different types of business and cannot fathom the risks of the other's business and thus not all of the risks of the project.

13.2 The Joint Venture: the formation of, and working within, a Joint Venture

When forming the Joint Venture ("JV") the parties must agree on various issues and principles under which the Joint Venture shall operate, such as:

1 Formation of the JV: object and purposes, administrative office;
2 Financial Plans: contributions of the members, warranties on contributions, profit and loss sharing and distribution, shared assets, termination of the JV;
3 Organisation and decision-making: general organisation, Steering Committee, Project Management;
4 Principles of joint operation: Good Faith and Fair Dealing; duty to inform and cooperate; exclusivity; same regard to the best interests of the Joint Venture as to individual interests of the JV Members; etc.;
5 Preparation, submission and acceptance of the JV's offer to the Owner;
6 Securities;
7 Insurance;

8 Financials and accounting, JV account, invoicing, payments, financing, unpaid working capital, accounts;
9 Taxes, tax obligations, value-added tax, taxation of the JV;
10 Confidentiality;
11 Liability: towards the Owner, towards the other JV Members, liability of the JV Project Management and limitation of liability; exclusion of further claims;
12 Expulsion: grounds for expulsion, procedure for expulsion, continuation of the JV, use of property, liability for work provided, liability for damages execution by substitution;
13 Duration and termination: Effective Date, causes for termination, consequences of termination;
14 Claims, dispute resolution and arbitration;
15 Choice of law.

Almost invariably at the beginning of the formation of every Joint Venture the prospective members start with aiming to preserve their own best interests, rather than focussing on the object and the purpose of the Joint Venture, acquiring and executing the project. The feeling always seems to be: "I must make sure the other(s) don't take advantage at my expense" in the assumption that the other(s) will try to do that. Trouble brews due to this approach, particularly if all prospective JV Members have that same attitude; it eats away at the foundation of the JV and the project, and if it isn't changed chances are high that the project will fail for the JV and its members.

Each prospective JV Member must focus on the success of the project: to complete it within the allotted time, within everyone's budget which allows for a fair reward for the Contractor, which meets the approval of all stakeholders in terms of execution and outcome, and which performs as required and has not given rise to disputes during construction.

This requires mutual trust between the JV Members, and to give trust requires both courage and honesty and an expressed decision by all members of the Supervisory Board of the JV that they and their companies will, without exception, relate to and work with each other that way. The JV Agreement should state this expressly and by signing it the members are committed. No member should allow any of its staff to act differently for any reason. This approach must be chosen from the very start; at the very first meeting regarding the formation of the JV this must be expressed by all representatives of the prospective members of the JV, and all must act accordingly from that moment on.

13.3 Financial Plans: contributions of the members, warranties on contributions, profit and loss sharing and distribution, shared assets, termination of the Joint Venture

The JV must establish its Financial Plans; what each of the members will contribute to the JV in working capital, in personnel, in equipment, etc.; each member's share in the JV will decide to what extent they will share in loss or in profit as the case may be at the end of the project. Also, their share in the JV will decide their interest in equipment bought by the JV in relation to the project for its execution. The percentage ratio of the value of each JV Member's contribution to the JV and the Works under the contract with the Owner ("the contract") decides the JV Member's share in the total of the JV and thus its share in profit or loss. Each member also must warrant its contribution in relation to the required performance of the contract and is to fully indemnify and hold harmless the other JV Members for any detrimental consequence of any grossly negligent or wilful breach of such warranty.

The JV Members also share an interest in equipment acquired by the JV for the execution of the Works under the contract proportionate to their contribution to the JV. At the end of the Works under the contract or if the Owner has terminated the contract for his convenience the value of the remaining assets of the JV shall be divided between the JV Members proportionate to their contribution to the JV.

13.4 Organisation and decision-making: general organisation, Steering Committee, Project Management

The bodies of the JV are the Steering Committee and the Project Management.

The Steering Committee is the decision-making body of the JV and it comprises authorised representatives of the respective JV Members, at least one of each JV Member, and the Steering Committee shall have the authority to make binding decisions. As continuity in the Steering Committee is best for the project the Steering Committee members should be appointed for the duration of the Works under the contract. Any change to be made in the membership of the Steering Committee must be notified as soon as is reasonably possible. The JV Members shall decide which Steering Committee member shall be the Chair of the Steering Committee.

Voting rights shall be proportionate to each JV Member's proportionate share in the JV but decisions shall be made with unanimous consent, particularly in JV's with two or three members who are familiar with each other. In larger JV's it may be helpful if the Steering Committee can decide to vote on a matter with a weighed majority if a unanimous consent cannot be reached, all in the interest of the project. But that can only happen if all members accept the possibility of being outvoted by a weighed majority of votes (e.g. ~ 75–80%) as the only other option would be to leave the JV or Consortium, as each of the members will be jointly and severally liable towards the Owner and if a member cannot agree fully with the bid to be sent to the Owner, the only option is to bail out of the JV or Consortium.

The Steering Committee appoints the Project Management or a single Project Manager where the project team can be managed by one person. In large projects it may be necessary to appoint a Project Management board, usually comprising a Finance Director, an Operations Director and a Project Director. It is the Project Management's duty to implement the decisions of the Steering Committee. As the Project Manager or the Project Director, as the case may be, is in almost daily contact with the Owner it makes sense to have the Project Manager or Project Director present at the Steering Committee meetings where decisions regarding the project will be made. It is the Project Manager's duty to inform the Steering Committee forthwith of any circumstance that is different than was foreseen if it is likely to have an impact on the cost of the Works or the time for completion under the contract and to advise the Steering Committee on how to address that circumstance.

It is of the utmost importance for the success of the project that all Steering Committee Members and the Project Management cooperate with one another in good faith.

13.5 Preparation, submission and acceptance of the Offer

One of the challenges of working in a Joint Venture or Consortium is aligning all the members with the requirements for successful bidding as set out in Chapter 4. This is an essential element as internal diversion on the elements of the bid to the Owner will make the bidding unsuccessful if not completely impossible (see Section 14.4).

On the Contractor's side the Bid Preparation team must comprise designers, estimators, engineers, a contract manager, a lawyer (preferably in-house) experienced in construction and construction contracts, dispute resolution and the use of Dispute Boards.

On the Owner's side there must be a similar multi-disciplinary team preparing the invitation to bid documents, evaluating the bids and preparing well-considered written advice to the Owner or the Owner's Supervisory Board if the Owner is a Joint Venture or a Consortium.

Primary focus of both the Contractor's team and the Owner's team must be the success of the project, not their own individual agendas and it is of vital importance that both Contractor's upper management and Owner's upper management stress this with the individual team members. The project is what is important, not long-term strategies, etc., though, of course, the project must fit into the company strategy of each JV or Consortium Member.

13.6 Securities

On the Contractor's side there is, most likely, an obligation to provide forms of securities: bonds and/or guarantees to be provided to the Owner. The cost of such bonds or guarantees to the Owner are to be borne by the Joint Venture and will be divided over the individual members in the proportion of their participation in the Joint Venture or Consortium. As, in many jurisdictions, the Joint

Venture or the Consortium is not a legal entity, the method for issuing guarantees will be case and jurisdiction specific.

13.7 Insurances

Insurance coverage is to be on terms and conditions as are reasonably available in the insurance market and that are customarily purchased by Contractors on similar projects with regard to size, technology and location. Any insurance proceeds under the Work Insurance are to be paid from the insurers to the Joint Venture account and are solely to be applied for the repair or rectification of any damage that has occurred. Deductibles, if any, are to be borne by the Joint Venture.

13.7.1 Marine Cargo/Transportation Insurance

Marine Cargo/Transportation Insurance, where applicable, is best maintained for not less than 110% of the Incoterms® (latest edition) CIF value to cover loss or damage to the Work during transportation through any means of transport from any warehouse anywhere in the world, until and including unloading at the site. For coverage one should refer oneself to the Institute Cargo Clauses (A) of the Institute of London Underwriters, provided war risks are available at base rate.

13.7.2 Construction/Erection All Risks Insurance

Construction/Erection All Risks Insurance must cover loss or damage to the Work on an all risks basis for not less than the full reinstatement cost, subject to sub-limits as are reasonably commercially available and with exclusions customarily required by the insurance market, such as the exclusion of the terrorism risk. This insurance must cover any site activity after unloading of the Work at the site, including storage on or near the site, construction, erection, assembly, cold and hot commissioning and testing until the Works are accepted or deemed to be accepted under the contract.

If Construction/Erection All Risks coverage is not taken out by the Owner, then unless otherwise agreed, the Joint Venture/Consortium must take out and maintain the insurance cover in the joint names of the Joint-Venture Members and the terms of insurance must provide for a waiver of recourse from the insurers against all insured parties and the cost of the cover shall be treated as cost to the Joint Venture.

13.7.3 Other insurance

Each Joint-Venture Member must take out and maintain at its own expense, and cause its subcontractors to take out and maintain at their own expense, the following insurance covers with reputable insurance carriers authorised to do business in the country of the site:

1. Employer's liability and workman's compensation insurance in accordance with any applicable law(s). If the law does not provide for any obligations in this respect or requires only some minimum limits, then the Joint-Venture Member must arrange this insurance in a manner and with limits as a prudent and reasonable Contractor in the same circumstances and environment would do. This insurance must be maintained in full force and effect during the whole time that Joint Venture's personnel are working in the execution of the Work on site.
2. Automobile liability insurance in the country of the site in accordance with any applicable law(s). If the law does not provide for any obligations in this respect or requires only some minimum limits, then the Joint-Venture Member must arrange this insurance in a manner and with limits as a prudent and reasonable Contractor in the same circumstances and environment would so do.
3. Comprehensive third-party liability insurance to cover the Joint-Venture Member's legal liability with a limit, unless otherwise agreed, of indemnity of not less than specified in the Joint Venture Agreement.
4. Joint-Venture Member's equipment insurance to cover loss or damage to Joint-Venture Member's equipment on an all risks basis for not less than the full replacement value, during any activity on the site and including delivery to the site.

The JV Members must provide each other with certificates of insurance regarding the relevant insurances shortly after such insurances are in place.

13.8 Financials and accounting, Joint Venture account, invoicing, payments, financing, unpaid working capital, accounts

13.8.1 Joint Venture account

The Project Management must be authorised on behalf of the Joint Venture and obliged to open and – together with one other Joint-Venture Member to be appointed by the Steering Committee – to operate a Joint Venture bank account. The Project Management must ensure that payments by the Employer under the contract are routed through the Joint Venture account. The authority to operate the account is to be revocable at any time by the Steering Committee.

The Steering Committee must formulate the instructions for operation of the bank account at its first meeting and as subsequently amended by the Steering Committee from time to time.

13.8.2 Invoicing

The Project Management is to invoice the Employer on behalf of the Joint-Venture Members to the extent permitted by the contract and by applicable tax regulations.

13.8.3 Payments

Any monies received on the Joint Venture account in excess of what is needed by the Joint Venture as working capital are to be distributed to the Joint-Venture Members pro rata their Proportionate Share unless otherwise decided by the Steering Committee.

All sums received by any Joint-Venture Member in connection with the contract shall promptly be paid into the Joint Venture Account and all monies to be paid to the Joint-Venture Members shall be paid from the Joint Venture account.

13.8.4 Financing

The Joint-Venture Members are likely to be required, in proportion to their respective proportionate participation, to provide guarantees to the bank(s) for the banking facilities or to provide such funds as the Steering Committee shall, from time to time, decide to be necessary in order to provide working capital for the performance of the contract in accordance with the Financial Plans. Working capital should not generate interest against the Joint Venture.

13.8.5 Unpaid working capital

If one of the Joint-Venture Members fails to pay its proportionate share of any sum or sums declared necessary to provide working capital, or agreed guarantee(s) within a certain period (e.g. 21 days after having received written notice to do so from the Steering Committee) such defaulting Joint-Venture Member should be liable to pay interest to the other Joint-Venture Member(s), on the amount of its proportionate share remaining unpaid or on the amount of its proportionate share of a guarantee or guarantees not provided. The amount of interest must be agreed in the JV Agreement. The payment of interest should be without prejudice to any rights under the JV Agreement of the Joint-Venture Member(s) not in default.

13.8.6 Accounts

The Project Management must keep the accounts of the Joint Venture in compliance with applicable laws and internationally accepted general accounting principles and according to the accrual method of accounting unless otherwise decided by the Steering Committee.

The books of accounts and records must be open for inspection of, and copy or extract by, the other Joint-Venture Members at all reasonable times.

The periodicity of accounting, issuance of profit and loss statement and balance sheet are to be determined by the Steering Committee in accordance with international standards and the respective requirements of the Joint-Venture Members.

Unless otherwise decided by the Steering Committee, the accounts of the Joint Venture are to be rendered by the Project Management to the Joint-Venture Members on a quarterly basis, with accounts for the last Gregorian calendar quarter being due no later than the end of the month following the respective quarter.

Unless otherwise decided by the Steering Committee, any objection by the Joint-Venture Members will have to be raised in writing no later than the end of the month following that of the rendition. If no objection has been raised within the specified time, the balance sheet and the profit and loss account are to be final and binding upon the Joint-Venture Members.

The Steering Committee is to appoint the Joint Venture's Independent Auditors and determine the scope of their mission.

13.9 Taxes

Each Joint-Venture Member must cooperate with the Project Management so that the Joint Venture complies with all tax regulations. In particular, each Joint-Venture Member must be responsible for supplying all such information necessary for preparing and submitting all necessary tax returns and tax payments, as legally required.

It is important to realise, particularly if the JV comprises companies from various jurisdictions, that where the Works take longer that a certain period, mostly three months, the JV will be seen by the tax authorities as having a "Permanent Establishment" making the JV and the individual JV Members liable for local corporate income tax apart from tax rules regarding value-added tax and similar consumption tax regulations which may be applicable right from the start of the project.

13.10 Confidentiality

13.10.1 Confidentiality requirements and exceptions

Each Joint-Venture Member must be obliged to use all business and technical information received from the other Joint-Venture Members in connection with the Joint Venture, the bid, the project and the Works, and which the disclosing Joint-Venture Member expressly has stated to be confidential or the confidential nature of which should be assumed on the basis of the circumstances of its disclosure or its contents, only for the purposes for which it was provided. He must treat it in the same way he expects his own business secrets to be treated and not make it available to third parties, unless the business or technical information in question:

1. Is generally available from public sources or in the public domain;
2. Is received at any time from any third party without a non-disclosure obligation to the disclosing Joint-Venture Member;
3. Is shown either to have been developed independently by the receiving Joint-Venture Member without reliance on the disclosing Joint-Venture

Member's confidential information or to have been known to the receiving Joint-Venture Member prior to its disclosure by the disclosing Joint-Venture Member; or

4 Must be disclosed to third parties for the purpose of performing this Agreement, provided such third parties are or become subject to an equivalent confidentiality obligation.

13.10.2 Confidentiality and the law

However, any Joint-Venture Member may be required and should therefore be allowed to disclose any confidential information to the extent that it is required to do so by the disclosure requirements of any law, rule or regulation or any order, decree, subpoena or ruling or other similar process of any court, tribunal, arbitral tribunal or governmental instrumentality or of any regulatory body having jurisdiction.

The ICC World Business Organisation has published a model Joint Venture Agreement, as well as a model Consortium Agreement[2], and these models reflect in detail the issues regarding confidentiality of information as well as other issues regarding working together in a Joint Venture or Consortium.

13.11 Liability

Either the contract with the Owner for the Works or the local law governing the contract will provide for the liability of the JV Members jointly and severally to the Owner for timely performance of the contract and to third parties in relation to the execution of the Works under the contract and all rights, obligations liabilities and risks are to be shared by the JV Members in proportion to their individual participation in the JV.

Towards each other the JV Members should only be liable where limitations of liability do not work: in case of gross negligence or wilful misconduct of one member towards the other(s). The Joint Venture Agreement must spell out in detail the liability issues of the JV Members towards each other.

13.12 Expulsion from the Joint Venture

13.12.1 Procedure of expulsion

Members of a Joint Venture should be able, in certain circumstances, to expel a member from the JV on the following grounds:

1 If, at any time whether before the award or during the contract, a Joint-Venture Member is being acquired by or is merging with an entity having a conflict of interest with the pursuit of the project; or

2 *ICC WBO*: "ICC Model Joint Venture Agreement and ICC Model Consortium Agreement". http://store.iccwbo.org/model-contracts

2 At any time whether before the award or during the contract if a Joint-Venture Member is disqualified by the Employer; or

3 If a Joint-Venture Member is lawfully excluded from the contract by the Employer; or

4 The Joint-Venture Member commits a material breach or persistent breaches of the Agreement, the contract or applicable laws and regulations which are irremediable or (if such breaches are remediable) where it fails to remedy the breaches without undue delay after having been given written notice by the Project Manager of it requiring such remedy. Such a breach might be in particular, but is not limited to, (i) a breach by a Joint-Venture Member of the undertakings under Article 5; or (ii) non-provision of the security stipulated in Article 7; or

5 A Joint-Venture Member ceases to hold any qualification, certification or registration required for the performance of its duties as a Joint-Venture Member, or becomes unfit to carry on its duties and obligations as a Joint-Venture Member; or

6 If, in the opinion of the Steering Committee, which shall be reasonable and justified, the behaviour of a Joint-Venture Member seriously endangers the successful performance of the Agreement and/or contract; or

7 A Joint-Venture Member has bankruptcy, composition or reorganisation proceedings or any other insolvency proceedings opened against it by court or other public authority; or has an order entered against it either appointing a receiver or trustee for, or issuing a levy attachment against a substantial portion of its assets, without this order being vacated, set aside or stayed within 60 days from the date of entry; voluntarily files a petition under the bankruptcy or equivalent insolvency law; becomes insolvent or faces a substantial deterioration (actual or imminent) in its assets; or has payment claims (existing under the terms of the Contract) or its membership interest in this Joint Venture seized by a creditor; consents to or applies for reorganisation under the bankruptcy or equivalent insolvency law; makes an assignment for the benefit of its creditors; or has involuntarily filed against it, a petition under the bankruptcy or equivalent insolvency law, which is not dismissed within 90 days after filing.

Expulsion must start with a decision adopted by the Steering Committee. The concerned Joint-Venture Member must be given the opportunity to defend itself, but its presence in the meeting should not be required and in no case will it have a voting right. Expulsion is to become effective on the day of the decision of the Steering Committee and it shall be binding if the Joint-Venture Member to be expelled has not, within 30 days of receipt of such decision, initiated Dispute Board proceedings in accordance with the JV Agreement or arbitration proceedings if the application of the Dispute Board has not been agreed upon. Expulsion cannot affect any of the existing rights or obligations of the Joint-Venture Member to be expelled or any claims against it.

The JV Agreement must contain provisions regarding continuation of the JV after expulsion of a member as well as the use of physical property (equipment, materials, etc.) and intellectual property against reasonable compensation if the JV cannot obtain a substitute at reasonable cost and in due time and the expelled JV Member is to facilitate this.

13.12.2 Liability for work provided

The expelled Joint-Venture Member(s) should not be relieved from any liability for or in connection with its performance of its contribution up to the date of expulsion, all subject to the limitations and/or exclusions of liability under the Agreement and/or the contract. The remuneration of the departing Joint-Venture Member for such supplies and services must be paid in accordance with the JV Agreement. The other Joint-Venture Members are to be entitled to require such payments to the departing Joint-Venture Member on such conditions as they reasonably think fit including provision of securities covering possible claims in accordance with the JV Agreement. In other words: the warranty for work performed by any JV Member survives expulsion or retirement from the Joint Venture.

13.12.3 Liability for damages

To the extent the expulsion is due to causes within the control of the expelled Joint-Venture Member the expelled member must be liable to the other Joint-Venture Members for damages (for example for the increase in cost to finalize the project and for compensation payable to the Employer for delay to the extent they result from the expulsion).

13.12.4 Expulsion by substitution

If expulsion is prohibited under the contract with the Owner, and therefore the earlier-mentioned provisions are not possible, then the remaining Joint-Venture Members should be able to decide in the Supervisory Board to reduce the contribution and, where appropriate, the liability of the concerned Joint-Venture Member(s) to 0 (zero) and remove its voting rights in the Joint Venture. Where legal and practical the consequences as described in Sections 14.12–14.14 must apply *mutatis mutandis* to expulsion by substitution.

13.13 Duration and termination of the Joint Venture

13.13.1 Effective date

The JV Agreement must state the date on which the JV Agreement becomes effective and that it shall continue until terminated in accordance with the JV Agreement.

13.13.2 Causes for termination

Upon a Joint-Venture Member's exercise of its right under mandatory law to terminate the Joint Venture, it shall leave the Joint Venture, and the remaining Joint-Venture Members shall continue the Joint Venture.

The Joint Venture shall be automatically terminated in the following events:

1 Withdrawal of the invitation to bid by the Employer or announcement by the Employer that it will not award the contract to any of the bidding parties; provided, that if the Employer announces within 180 days of such withdrawal or announcement of non-award that it will re-bid the project, then this Agreement shall remain in effect for the re-bid unless otherwise agreed by the Joint-Venture Members; or
2 The expiry of the period of validity of the Offer unless the Joint-Venture Members have agreed in writing to prolong the validity of the JV Agreement beyond such period; or
3 If the final Offer of the Joint-Venture Members is not accepted and no further negotiations are carried out between the Employer and the Joint Venture; or
4 When the Joint-Venture Members have unanimously so decided, and recorded such decision in writing; or
5 Failure of the Joint-Venture Members to agree on an Offer after negotiating in good faith for a reasonable time, provided that jointly two Joint-Venture Members or the Joint Venture Project Management give the other Members e.g. five days written notice of the intention to terminate the Agreement; or
6 In the event that the contract is awarded to the Joint Venture, once the contract has been performed or terminated and the Joint-Venture Members have fulfilled all their obligations and received their rights in accordance with the contract and the JV Agreement.

The termination of the JV Agreement should not affect any of the rights and obligations of the Joint-Venture Members accrued up to the date of the event causing the termination of the Agreement, including in respect of the sharing of profit and loss.

13.14 Claims, dispute resolution and arbitration

The JV Agreement must provide that all disagreements, claims and disputes are to be resolved solely by the procedures set out in the JV Agreement, being first an attempt to settle the issues between them through amicable settlement and should that such attempt fail through an appointed Dispute Board (it is advisable to appoint such board under the ICC Dispute Board Rules, as these rules provide for a robust arrangement of dispute resolution which is internationally acclaimed as they provide a well-balanced set of rules also obtainable via the ICC WBO website).

Should any member of the JV fundamentally disagree with the Dispute Board's decision, the third tier of dispute resolution should be arbitration under the ICC Arbitration Rules, again as these rules provide a balanced set of rules for arbitration and the ICC also provides a Court of Arbitration where the dispute can be tried as opposed to e.g. the UNCITRAL Rules which do not provide for a Court of Arbitration.

13.15 Choice of law

It is wise to choose the law applicable to the JV Agreement and any disputes that may occur in relation to the laws applicable in the country where the site under the contract with the Owner is located, particularly if the members of the JV come from various jurisdictions. If they all come from one jurisdiction it might make sense to choose the laws of their common jurisdiction to apply to the JV Agreement, albeit that they will be subject to the laws of the jurisdiction where the site is located and the Works are executed with regard to the execution of the contract.

13.16 The Consortium: the formation of, and working within, a Consortium

As stated in Section 13.1 of this chapter, parties form a Consortium rather than a Joint Venture when they are rooted in different types of business and cannot fathom the risks of the other's business and thus not all of the risks of the project. Where within the Joint Venture claims on the other JV Members are excluded unless damages were caused by gross negligence or wilful misconduct, such an exclusion cannot be the case if the individual members cannot assess the risks in the scope of work of the other members, as they cannot influence the daily execution of work by the other parties due to the absence of knowledge necessary to assess and manage all the risks of the daily works.

The Introduction to the ICC Model Consortium Agreement explains the difference between a Joint Venture and a Consortium as follows:

> The main difference between the two is the extent to which the partners for their own immediate benefit share performance risks, i.e. the risks innate to the nature of their resources, technologies and processes. In a joint venture the members share all risks, liabilities, rights, benefits and profits in proportion to their participation in the joint venture, such proportion normally being determined based on the resources that each member will supply to the joint venture for the execution of the total scope of work under the contract with their employer, to whom they are jointly and severally liable for all the contractual obligations. Joint venture members are generally kindred spirits in their field of business and understand each other's business and the risks therein.
>
> In a consortium the co-operating parties, though jointly and severally liable for the whole of the contractual obligations towards their employer, are, towards the other consortium members, liable for the proper execution of only their individual

> parts of the total scope of work. This form of co-operation should be chosen if the consortium members are not familiar with the intricacies of the other members' business and risks therein and therefore have no means to control either the probability of the risk event materializing or the impact that the occurrence of the risk event may have. Consortium members are often operating in industries adjacent to each other, e.g. civil contractors and engineering contractors or dredging contractors.

So, in a Consortium each member has his own scope of work, taken from the total scope or work from the contract with the Employer and each member is responsible for his own scope of work and the successful completion thereof and therefore there is no sharing of profit or loss in a Consortium, each member derives his profit or loss, as the case may be, from his own scope of work.

Depending on the (tax) laws of the country where the site is located, each individual Consortium Member invoices the Employer directly, or, if the laws do not allow that, a Consortium Member sends his statement of account of his work to the Consortium Leader who will subsequently invoice the Employer on behalf of the individual Consortium Member.

The Consortium Members each nominate an authorised member to the Steering Committee of the Consortium. The Consortium Members through the Steering Committee appoint the Consortium Leader whose duties are:

1 To provide technical, commercial and organisational coordination of the Consortium Members in the bidding phase and during the performance of the contract;
2 To act as the spokesperson for the Consortium in negotiations with the Employer, authorities or any third parties which are of joint interest, including preparing the necessary correspondence;
3 To submit the bid for the Consortium and coordinate necessary formalities, in particular regarding a potential mandatory registration of the Consortium, to keep the necessary books, and to prepare and submit tax returns for the Consortium; to convene, preside and keep records of the meetings of the executive body(ies) of the Consortium;
4 To make proposals for joint insurance coverage of all the Consortium Members pursuant to the Consortium Agreement;
5 To coordinate the establishment of a joint construction site;
6 To coordinate the preparation of progress reports and other documentation to be submitted to the Employer;
7 If and as required, to keep an integrated information and digital documentation system and organise the sharing of such information and documentation with the Consortium Members and with the Employer as required;
8 To collect data for invoicing the Employer and to collect payments from the Employer to the extent and in the way foreseen in the Consortium Agreement;
9 To perform any other duties assigned to the Consortium Leader by the Consortium Agreement or by a respective resolution or decision of the

Steering Committee or of the Consortium Members supplementing or interpreting the Consortium Agreement.

It is common that the Consortium Leader, from the moment the contract with the Employer is in effect, receives a fee in consideration for the execution of his duties, agreed as a lump sum between the Consortium Members in advance.

13.17 Takeaways

Forming a Joint Venture usually turns out to be not such an easy undertaking. Parties forming a JV often are afraid that the other member(s) will take advantage of them and they respond very often by engaging in the same tactics: try to outsmart one another.

At the very beginning the members have to decide what they will contribute to the JV in capital, personnel and equipment; what the reward for that contribution will be; their warranties on these contributions; how profit and loss will be shared; and how the termination procedure is to be at the end.

They need to decide where the administrative office will be (often on or near the site), that the supreme power in the JV is vested in the Steering Committee comprising a delegate from every member, who the Project Manager will be, etc. Decisions are to be made unanimously and if a member cannot agree with the other members he must step out of the JV.

They must agree in their JV Agreement to the principles of joint operation which are to be Good Faith and Fair Dealing: duty to inform the other members, to operate in exclusivity regarding the bid to the Owner and to guard the best interests of the JV as they do their own interests.

The more contribution to the JV the more profit a member receives but also the more risk he is exposed to. Everything a member of a JV is entitled to or exposed to in a JV is in proportionate to its participation in the JV which is relative to its contribution to the JV. Towards the Owner the JV Members are jointly and severally liable.

The JV is to have a Joint Venture bank account to which all monies receivable flow and from which all dues of the JV are paid. When a JV Member does not pay its contribution in time he may be charged interest on the unpaid contribution. The Steering Committee may decide to expel a derelict member of the JV and the consequences of such expulsion.

The reason for parties to cooperate in a Consortium is that they are each rooted in a different type of business but given the size and the risk of a particular project would like to undertake the project together. To organise such cooperation, parties form a Consortium. The power structure of the Consortium is similar to the Joint Venture; the supreme power is vested in a Steering Committee comprising delegates from the members of the Consortium. The difference is that the individual members of the Consortium will each have their own scope of work, where in a Joint Venture the scope of work is shared between the members. In a Consortium the members will have difficulties attempting to assess the risks in the other member's scope of work and how that member is dealing with those risks.

Towards the Owner the Consortium Members are jointly and severally liable.

CHAPTER 14

The power structures of a Joint Venture and a Consortium and the decision-making process

14.1 The Joint Venture: strategic management

Companies in construction that want to do well need to employ strategy to gain the reputation of reliability to provide services to and beyond expectations of its Owners; to improve and innovate continuously; to be an attractive employer for employees so the company can attract the best people in their trade; to be a solid investment for shareholders by meeting their expectations; and to grow. In the previous chapter the reasons for companies to join forces have been listed and explained.

For the duration of the project, the Joint Venture or Consortium, as the case may be, is like an unincorporated company, it needs a solid management structure.

14.2 The management structure of the Joint Venture

14.2.1 The Steering Committee

The supreme power of the JV is to be vested in the Steering Committee comprising at least one representative of each of the JV Members, and these representatives shall have the right to legally bind the respective JV Member they represent. The Steering Committee is in charge of:

- The strategic management of the JV and of the communication with the Owner in relation to the construction contract and the negotiations of the changes to that contract or termination of it;
- Setting the overall Project Coordination Procedure and control of the Project Management and all specific matters it must decide on as per the JV agreement.

If the JV Members decide it is necessary to have more than one Steering Committee representative per JV Member, say if the members want more than one discipline represented in the Steering Committee, then each of the JV Members shall appoint an equal number of Steering Committee representatives, each fully empowered to bind the JV Member it represents. The JV Members shall bear the costs and expenses of their own Steering Committee member(s).

14.2.1.1 Decision-making process in the Steering Committee

The Steering Committee decides by unanimous consent; it cannot be that one member of the JV is forced by the others to accept something they are very much against, that is unless the JV Agreement provides for a majority decision for certain issues. That could be conceivable in a Consortium, where some of the members have a good understanding of a certain topic and some others do not. As there still will be a joint and several liability for all members it could be a good idea, in some cases, that those in favour of a certain decision provide an indemnity to the member(s) not in favour of the decision. Imagine that an Owner wishes to issue a variation order that brings quite some risk with it but the members who are in favour are sure that they can manage that risk successfully and one or two other members are against executing such a variation as they understand the risk but do not have the knowledge necessary to manage that risk. Those in favour of the variation could decide to indemnify those against it for any negative consequences of the execution of the variation desired by the Owner; all in the interest of a successful project. In a Joint Venture however there can only be room for strict joint liability.

Needless to say, the Chair of the Steering Committee must record decisions made and keep minutes of meetings and circulate such records to the other members.

Each JV Member delegates an authorised representative to the JV who shall have the authority to bind the respective Joint-Venture Member by whom they have been appointed, for the purposes of this Joint Venture Agreement. It is sensible to also appoint an alternate Steering Committee member for the occasion that the appointed authorised representative is not able to attend a Steering Committee meeting in which case the appointed alternate may take his place, having the same authority to represent the JV Member that appointed him. The appointment of an alternate also should prevent a member from intentionally staying away from a Steering Committee meeting in order to block a decision from the Steering Committee that the Member does not agree with. These things do happen in practice unfortunately and when they do the project suffers.

In the event of a deadlock at a Steering Committee meeting, the meeting is to be adjourned and to reconvene several days later. In the event of a continued deadlock, the matter must forthwith be referred, in writing, by the Chair of the Steering Committee, to the chief executive officers of each of the Joint-Venture Members or to another person of suitable standing and authority chosen by the respective Joint-Venture Members, who then must meet within two or three weeks of the matter referred to them. They must work together to reach a unanimous decision which is to be binding on all Joint-Venture Members, failing which the disagreement must be referred to the agreed dispute resolution process.

14.2.2 The Project Management

The Project Management for smaller projects a one-person job, assisted by others, for the larger projects usually a group of people comprising a "Project

Director", a finance manager and an operations manager, so a three headed body, or a one-person lead, runs the day-to-day business of the JV, subject to the overriding authority of the Steering Committee. Usually the group of people is appointed from the personnel of the members of the JV on the basis of the best people available for the project, including the criterion: "Can they work well together?"

It is essential that the Steering Committee and the Project Management have regular contact, inform each other and be open to any communication, good news or bad news, so that early measures can be initiated to avoid cost overruns and delays in the project.

14.3 Tasks of the Project Management in a Joint Venture

The tasks of the Project Management are:

1 Implementing the decisions of the Steering Committee;
2 Opening the joint bank account(s); such bank account(s) must always be operated together with one other Joint-Venture Member which is to be appointed by the Steering Committee;
3 Representation of the Joint Venture in its relations with the Employer and other third parties, as well as entering into contracts, such as sub-contracts on behalf of the Joint Venture as and when authorised by the Steering Committee under a general or specific authorisation;
4 The purchase of any equipment and/or materials required to perform the Works upon specific authorisation of the Steering Committee, where these are not provided by the Joint-Venture Members. Upon completion of the Works or at any earlier time decided by the Project Management in conjunction with the Steering Committee, such equipment and/or materials may be sold by the Project Management;
5 Providing technical, commercial and organisational coordination of the Joint-Venture Members in the bidding phase and during the performance of the contract;
6 To the extent applicable, submission of the Offer for the Joint Venture and coordination of the necessary formalities, in particular regarding a potential mandatory registration of the Joint Venture;
7 Keeping the necessary accounting books, and preparing and submitting tax returns for the Joint Venture as applicable;
8 Making proposals for joint insurance coverage of all the Joint-Venture Members;
9 Coordinating the establishment of the joint construction site(s);
10 Coordinating the preparation of progress reports and other documentation to be submitted to the Employer and generally keeping records of meetings with the Employer and/or third parties in relation to the project;

11 In the event of changes with respect to the Works that are required by the Employer, coordinating the preparation of proposal for the approval of the Steering Committee and upon approval by the Steering Committee to negotiate with the Employer;

12 If and as required, keeping an integrated information and digital documentation system and organising the sharing of such information and documentation with the Joint-Venture Members and with the Employer as required; and collecting data for invoicing the Employer and to collect payments from the Employer. Performing any other duties assigned to the Project Management through the Joint Venture Agreement or through a decision of the Steering Committee having such effect.

14.4 The contributions of members to the Joint Venture

"Contribution" is what a Joint-Venture Member brings into the Joint Venture that has monetary value such as tangible and intangible assets including know-how and intellectual property, services, material, equipment and other resources including cash resources. Needless to say that contribution is a frequent source of heated debate regarding the monetary value of the contributions, during the formation of the Joint Venture. The best way to handle this issue is to make a list of the contributions each member is willing and able to make and at what monetary value, and a list of intellectual property, services, material, equipment and other resources needed for the project and have the Steering Committee decide which JV Member will contribute which intellectual property, services, material, equipment and other resources and at what monetary value. Such a list with most items on it was most likely prepared in the bidding stage when preparing the Offer to the Owner. The reason for the heated debate is mostly that companies are considering their own best interest when handling these issues as opposed to the project's best interest. When the members act in accordance with the principles of Good Faith and Fair Dealing a heated debate can be avoided. It is in each JV Member's best interest to focus on achieving the project's objectives.

The Joint-Venture Members make the contributions to the JV by way of granting the JV a right of use and, as the case may be, right of disposal or consumption of such contributions for the needs and duration of the Joint Venture Agreement and to the extent required by the contract. The nature and extent of such contributions may be amended from time to time and any resulting consequences (for example relating to the Proportionate Values) do require a unanimous decision of the Steering Committee.

Each Joint-Venture Member must:

- Warrant that its respective contribution fully complies with the requirements of the contract related to the scope of such contribution; and
- Fully indemnify and hold harmless the other Joint-Venture Members for any detrimental consequence to the Joint Venture of any grossly negligent or wilful breach of such warranty.

14.5 Profit and loss sharing in the Joint Venture

Joint-Venture Members share, or bear, the rights, obligations, liabilities, risks and all net profits or net losses of their joint performance of the contract in proportion to the respective Proportionate Value of their participation in the JV.

When the project is finished and the Owner has paid all monies due to it under the contract and on realisation of all assets of the Joint Venture and the settlement of all its outstanding obligations and liabilities, the Steering Committee must cause a final account to be prepared, showing the total net profit earned or total net loss incurred by the Joint Venture. Once the Steering Committee has agreed on the final account, each of the Joint-Venture Members must receive its proportionate share of the net profit, or pay its proportionate share of the net loss of the JV, after taking into account any interim distribution on account of profit.

Needless to say, the Steering Committee must seek expert tax advice on what the relevant tax legislation regulates with regard to the method of recognizing profit or loss.

Additionally, tax jurisdictions sometimes regulate that when the project takes longer than a certain number of months for construction the Joint Venture is considered to have a "Permanent Establishment" which makes the JV subject to income tax.

The total net profit or net loss is the sum remaining or outstanding after deducting from the total of amounts received (excluding all working capital contributions) by the Joint Venture of all costs properly incurred including:

1 The cost of any special service performed by any of the Joint-Venture Members at the prior request of the Steering Committee;
2 The costs of travelling and out-of-pocket expenses (but not the salaries) incurred by members of the Head Office staff of either of the Joint-Venture Members not engaged full time on the contract, in visiting the site in the interest of the execution of the contract;
3 All rates, taxes or duties levied on the Joint Venture;
4 Fees expressly agreed by the Steering Committee prior to their incurrence, which are paid to third parties in connection with the preparation of the Offer and the performance of the contract and the settlement of claims arising under the contract.

14.6 Principles of joint operation

In carrying out their obligations the Joint-Venture Members must act in accordance with the principles of Good Faith and Fair Dealing. The provisions of the Joint Venture Agreement, as well as any statements made by the Joint-Venture Members in connection with it, must be interpreted in accordance with the principles of Good Faith and Fair Dealing.

14.6.1 Duty to inform and cooperate

Each Joint-Venture Member must promptly inform the other Joint-Venture Members of any matters of which it becomes aware and which may materially affect the preparation of the Offer, performance of the contract or the execution of the Works or, in any way, reasonably be required or necessary for the other Joint-Venture Members to prepare the Offer, perform the contract and execute the Works.

Each Joint-Venture Member must cooperate in the joint performance of the contract in accordance with the Joint Venture Agreement.

14.6.2 Exclusivity

The Joint-Venture Members must work together on an exclusive basis to achieve the objectives of the JV. In particular, no Joint-Venture Member should by itself or with other parties either make Offers or enter into agreements concerning the project or any part thereof.

This exclusivity obligation applies also to the contribution to the Joint Venture. This exclusivity obligation should apply until the Joint Venture is wound up pursuant to the JV Agreement. If a Joint-Venture Member voluntarily leaves or is expelled from the Joint Venture for any reason for which it is responsible, the exclusivity obligation must bind such Member until the Joint Venture is wound up.

Throughout the life of the Joint Venture Agreement, each Joint-Venture Member must act in a manner consistent with the agreed common purpose and best interests of the Joint Venture and he must have the same regard to the legitimate expectations of the other Joint-Venture Members in respect of the Joint Venture as for its own interests therein.

14.7 Preparation, submission and acceptance of the Offer

The Joint-Venture Members must jointly prepare and submit the Offer for the execution of the project to the Owner. That Offer must be in the form required by the Owner and with qualifications and changes to the Owner's terms and conditions as the Joint-Venture Members shall mutually agree, and after its preparation and submission, the Joint-Venture Members shall be jointly and severally bound by the provisions of the Offer and none of them may seek to vary them without the previous written consent of the others.

The Joint-Venture Members prepare the Offer in accordance with a timetable, agenda and programme to be established as soon as possible after the signing of the JV Agreement.

In proportion to their participation in the JV, the Joint-Venture Members share, as closely as practical, in all work required for the preparation of the proposal and make services available from their own resources wherever possible. Joint-Venture Members may not engage external services without the (prior written) agreement of the other Joint-Venture Member(s). Each Joint-Venture Member bears its own costs and expenses incurred in preparing the proposal, whether the proposal and/or negotiations are successful or not. External services appointed by prior written agreement between the Joint-Venture Members shall be shared in the proportion to the participation of the JV Members.

Withholding of approval of the Offer by any of the Joint-Venture Members cannot be subject to arbitration, but will automatically terminate the JV Agreement unless the JV Agreement contains a "survival clause", which provides that the JV shall continue if there will be a new call for Offers by the Owner for the project within 12 months after the Owner cancelled the project or rejected the Offers made by the various bidders. Such survival clause one finds regularly in JV Agreements and is incorporated in the JV Agreement to prevent "leaking" of information to other parties by a former JV Member who engages in subsequent bidding for the project in a different Joint Venture than for the original bid.

Any bonds, guarantees or indemnities required by or arising out of the terms and conditions of the Offer are to be arranged by and to be paid for by the Joint-Venture Members in the proportions of their financial interests in the JV.

None of the Joint-Venture Members should in any way act either alone, or jointly with, by or through any other party in a manner likely to be detrimental to the Offer or cause loss or injury to the Joint Venture.

While the Joint Venture is in existence, a Joint-Venture Member is not allowed to participate in any other Offer for the project, either directly or through any subsidiary or other associated company or otherwise without the prior written consent of the other Joint-Venture Members.

14.8 Securities

Any and all securities, bonds or guarantees, required by or arising out of the terms and conditions of the contract, are to be provided by the Joint Venture and the cost of such bonds, securities guarantees or counter-guarantees must be borne by the Joint Venture. Any and all counter-guarantees required are to be issued by the Joint-Venture Members within the agreed number of days of receipt of notice from the Steering Committee requiring them to do so. Because an unincorporated Joint-Venture is not treated as a legal entity in many jurisdictions, the method for issuing guarantees, bonds or other securities will be case and jurisdiction specific and the members should seek legal advice on this issue.

14.9 Insurances

14.9.1 Work Insurance

14.9.1.1 General principles

Any insurance cover should be on terms and conditions as they are reasonably available in the insurance market and that are customarily purchased by Contractors on similar projects with regard to size, technology and location. Any insurance proceeds under the Work Insurance must be paid by the insurers to the Joint Venture account and are to be applied for the repair or rectification of any damage that has occurred. Deductibles, if any, must be borne by the Joint-Venture Members in proportion to their participation in the JV.

14.9.1.2 Marine Cargo/Transportation Insurance

Marine Cargo/Transportation Insurance, where applicable, should be maintained for not less than 110% of the Incoterms® (latest edition) CIF value to cover loss or damage to the Work during transportation with conveyances of whatsoever kind from any warehouse worldwide until and including unloading at the site. This cover should not be less than the internationally known Institute Cargo Clauses (A) of the Institute of London Underwriters, provided war risks are available at base rate and the deductible will not be higher than the amount per occurrence as agreed between the JV Members.

14.9.1.3 Construction/Erection All Risks Insurance

Construction/Erection All Risks Insurance must cover loss or damage to the Work on an all risks basis for not less than the full reinstatement cost, subject to sub-limits as are reasonably commercially available and with exclusions customarily required by the insurance market, such as the exclusion of the terrorism risk. This insurance shall cover any site activity after unloading of the Work at the site, including storage on or near the site, construction, erection, assembly, cold and hot commissioning and testing until the Works are accepted or deemed to be accepted under the contract.

If not taken out by the Employer, then unless otherwise agreed, the Project Management shall effect and maintain the insurance cover in the joint names of the Joint-Venture Members and shall provide for a waiver of recourse from the insurers against all insured parties, in which case the cost of the cover shall be treated as cost to the Joint Venture.

14.9.1.4 Other insurance

Each Joint-Venture Member must take out and maintain at its own expense, or shall cause its subcontractors to effect and maintain at their own expense, the following insurance cover with reputable insurance carriers authorised to do business in the country of the site:

Employer's liability and workman's compensation insurance in accordance with any applicable law(s). If the law does not provide for any obligations in this respect or requires only some minimum limits, then the Joint-Venture Member shall arrange this insurance in a manner and with limits as a prudent and reasonable Contractor in the same circumstances and environment would so do. This insurance shall be maintained in full force and effect during the whole time that Joint Venture personnel are working in the execution of the Work on site.

Automobile liability insurance in the country of the site in accordance with any applicable law(s). If the law does not provide for any obligations in this respect or requires only some minimum limits, then the Joint-Venture Member shall arrange this insurance in a manner and with limits as a prudent and reasonable Contractor in the same circumstances and environment would so do.

Comprehensive third-party liability insurance to cover the Joint-Venture Member's legal liability with a limit, unless otherwise agreed, of indemnity of not less than specified by the JV Agreement.

Joint-Venture Member's equipment insurance to cover loss or damage to Joint-Venture Member's equipment on an all risks basis for not less than the full replacement value, during any activity on the site and including delivery to the site.

On request of another Joint-Venture Member, a Joint-Venture Member shall provide evidence to the other Joint-Venture Members that the Joint-Venture Member's other insurance is in place as soon as possible after such request.

14.10 Financials and accounting

14.10.1 Joint Venture account

On behalf of the Joint Venture, the Project Management, where required, is authorised and obliged to open and – together with one other Joint-Venture Member who shall be appointed by the Steering Committee – to operate a Joint Venture bank account. The Project Management shall use best efforts to ensure that payments by the Employer under the contract are done through the Joint Venture account. The authority to operate the account shall be revocable at any time by the Steering Committee.

The bank account is to be operated in accordance with the instructions laid down by the Steering Committee at its first meeting or as subsequently amended by the Steering Committee from time to time.

14.10.2 Invoicing

The Project Management invoices the Employer on behalf of the Joint-Venture Members to the extent permitted by the contract and by applicable tax regulations.

14.10.3 Payments

Any monies received on the Joint Venture account in excess of what is needed by the Joint Venture as working capital shall be distributed to the Joint-Venture

Members pro rata their Proportionate Value unless otherwise decided by the Steering Committee.

All sums received by any Joint-Venture Member in connection with the contract shall promptly be paid into the Joint Venture Account and all monies to be paid to the Joint-Venture Members shall be paid from the Joint Venture account.

14.10.4 Financing

The Joint-Venture Members must, in proportion to their respective Proportionate Value, provide guarantees to the bank(s) for the banking facilities or must provide such funds as the Steering Committee, from time to time, decides to be necessary in order to provide working capital for the performance of the contract in accordance with the Financial Plans. Working capital should not earn interest against the Joint Venture.

14.10.5 Unpaid working capital

Where one of the Joint-Venture Members fails to pay its proportionate share of any sum or sums declared necessary to provide working capital, or agreed guarantee(s) within an agreed period after having received written notice to do so from the Steering Committee, the defaulting Joint-Venture Member is to be liable to pay interest to the other Joint-Venture Member(s) on the amount of its proportionate share remaining unpaid or on the amount of its proportionate share of a guarantee or guarantees not provided. The payment of interest should be without prejudice to any rights of the Joint-Venture Member(s) not in default under the JV Agreement.

For the event that a defaulting Joint-Venture Member has not made payment of its proportionate share within the period stated in the JV Agreement, the JV Agreement must provide that such JV Member is not allowed to pay any further sum to provide working capital during the course of the contract, except to the extent previously approved in writing by the other Joint-Venture Member(s) and the respective shares of the Joint-Venture Members in the profits calculated in the Offer to the Owner are to be adjusted and if necessary re-adjusted pro rata the sum or sums provided by each of the Joint-Venture Member(s) as working capital, or as a guarantee or guarantees. However, the proportionate liability of the defaulting Joint-Venture Member for losses, costs, expenses risks and obligations must remain unchanged.

So long as a Joint-Venture Member remains in default regarding these payments, its Authorized Representatives may not vote at the meetings of the Steering Committee and the decisions of the Steering Committee are to be validly taken and binding, also upon the Joint-Venture Member in default, by the exercise of the voting rights of the Joint-Venture Members not in default.

14.10.6 Accounts

The Project Management must keep the accounts of the Joint Venture in compliance with applicable laws and internationally accepted general accounting principles and according to the accrual method of accounting unless specified in more detail in the JV Agreement or as otherwise decided by the Steering Committee.

The books of accounts and records must be open for inspection of and copy or extract by the other Joint-Venture Members at all reasonable times.

The periodicity of accounting, issuance of profit and loss statement and balance sheet are to be determined by the Steering Committee in accordance with international standards and the respective requirements of the Joint-Venture Members.

The accounts of the Joint Venture must be rendered by the Project Management to the Joint-Venture Members on a quarterly basis unless otherwise decided by the Steering Committee. The accounts for the last Gregorian calendar quarter must be due no later than the end of the month following that respective quarter.

The Steering Committee is to appoint the Joint Venture's Independent Auditors and to determine the scope of their mission.

14.10.7 Taxes

14.10.7.1 Tax obligations

The Joint-Venture Members must cooperate with the Project Management so that the Joint Venture will comply with all applicable tax regulations. In particular, each Joint-Venture Member will be responsible for supplying all necessary information for preparing and submitting all necessary tax returns and tax payments, as legally required.

14.10.7.2 Value-added tax

If the Joint Venture is subject to sales or value-added tax, the relevant tax obligations are to be fulfilled by the Project Management on behalf of the Joint Venture.

Details of necessary procedures (with regard to invoicing procedure, tax declarations, documents and respective tax payments) are be agreed upon separately by the Steering Committee.

14.10.7.3 Taxation of the Joint Venture

In the event that the tax authorities tax the Joint Venture on the basis of income from the entire contract with the Owner, the Joint-Venture Members must closely cooperate and agree in the Steering Committee on a joint approach in responding to the tax authorities' position. If the Joint Venture is taxable, the Joint Venture Project Management will coordinate the tax filing and tax payment process.

Taxes other than corporate income tax assessed on the Joint Venture shall be borne by the Joint-Venture Members pursuant to their Proportionate Values.

14.11 Confidentiality

Each Joint-Venture Member must use all business and technical information received from the other Joint-Venture Member(s) in connection with the JV Agreement, the

Offer, the project and the Works, and which the disclosing Joint-Venture Member expressly states to be confidential or of which the confidential nature is reasonably obvious solely for the purposes for which it was provided and must treat it in the same way as its own business secrets and not make it available to third parties, unless the business or technical information in question:

1 Is generally available from public sources or in the public domain;
2 Is received at any time from any third party without a non-disclosure obligation to the disclosing Joint-Venture Member;
3 Is shown either to have been developed independently by the receiving Joint-Venture Member without reliance on the disclosing Joint-Venture Member's confidential information or to have been known to the receiving Joint-Venture Member prior to its disclosure by the disclosing Joint-Venture Member; or
4 Must be disclosed to third parties for the purpose of performing this contract, provided such third parties are or become subject to an equivalent confidentiality obligation.

A Joint-Venture Member must be allowed to disclose any confidential information if, and to the extent that, it is required to do so by the disclosure requirements of any law, rule or regulation or any order, decree, subpoena or ruling or other similar process of any court, tribunal, arbitral tribunal or governmental instrumentality or of any regulatory body having jurisdiction. The prudent action is, of course, to inform the Joint-Venture Member that initially provided the information (the "owner of the information") that one is required to disclose the confidential information and to cooperate with the owner of the information to obtain proprietary or confidential treatment for such confidential information by the third party to whom the confidential information is disclosed or to seek protective orders limiting the dissemination and use of the confidential information.

This confidentiality obligation must also apply to the contents of the Joint Venture Agreement.

The confidentiality obligation created in the JV Agreement is to continue for a period after termination of the Joint Venture the length of which is to be decided by the JV Members in the JV Agreement.

Any Publications in the media or of any other kind by a Joint-Venture Member or initiated by a Joint-Venture Member referring to the Works must require the prior written approval of the other Joint-Venture Members, which approval should not be unreasonably withheld.

14.12 Liability

14.12.1 Liability towards the Employer and/or third parties

The Joint-Venture Members are jointly and severally liable to the Employer for timely performance of the contract and to third parties in relation to the execution of the project either by a provision in the JV Agreement or by applicable law. All rights, obligations, liabilities and risks will be shared between the Joint-Venture Members in accordance with the Proportionate Values of their participation in the JV and set out in the JV Agreement.

14.12.2 Liability towards the other Joint Venture Members

Towards each other the Joint-Venture Members should only be liable in the event of gross negligence or wilful misconduct by any Joint-Venture Member causing a loss to the Joint Venture, except as otherwise agreed in the JV Agreement.

14.12.3 Liability of the Joint Venture Project Management and limitation of liability

The Joint Venture Project Management is to be liable for damage caused to any and all of the Joint-Venture Members through negligence with respect to the performance of its duties as described in the JV Agreement. Such liability should, in aggregate, be limited to the amount of the fee of the Joint Venture Project Management. The Joint Venture Project Management should never be liable for loss of profit, loss of use, loss of data or information, loss of contracts or business opportunities or any punitive damages to the extent consistent with mandatory law and regardless of whether the liability claim is based on breach of contract, breach of warranty, negligence, strict liability, tort or any other legal theory and should also apply for the benefit of the employees of the Joint Venture Project Management. Any excess damage is to be divided among the affected Joint-Venture Members in proportion to their participation in the Joint Venture.

14.12.4 Exclusion of further claims

Amongst each other, the Joint-Venture Members should have no claims for damages and cost reimbursement other than those set forth in the JV Agreement to the extent consistent with mandatory law.

14.13 Expulsion

14.13.1 Grounds for expulsion

Unless prohibited under the contract with the Owner, there should be a provision in the JV Agreement that a Joint-Venture Member may be expelled from the Joint Venture if the other Joint-Venture Member(s) cannot reasonably continue the Joint Venture with a particular Joint-Venture Member. Relevant circumstances giving rise to such expulsion could be:

1 At any time whether before the award or during the contract a Joint-Venture Member being acquired by or merged with an entity having a conflict of interest with the pursuit of the project; or

2 At any time whether before the award or during the contract if a Joint-Venture Member is disqualified by the Employer; or

3 If a Joint-Venture Member is lawfully excluded from the contract by the Employer; or

4 The Joint-Venture Member commits a material breach or persistent breaches of the Agreement, the contract or applicable laws and regulations which is irremediable or (if such breach is remediable) where it fails to remedy that breach without undue delay after having been given written notice by the Project Manager of it requiring such remedy.

5 A Joint-Venture Member ceases to hold any qualification, certification or registration required for the performance of its duties as a Joint-Venture Member, or becomes unfit to carry on his duties and obligations as a Joint-Venture Member; or

6 If, in the opinion of the Steering Committee, which shall be reasonable and justified, the behaviour of a Joint-Venture Member seriously endangers the successful performance of the Agreement and/or contract; or

7 A Joint-Venture Member has bankruptcy, composition or reorganisation proceedings or any other insolvency proceedings opened against it by court or other public authority; or has an order entered against it either appointing a receiver or trustee for, or issuing a levy attachment against a substantial portion of its assets. NOTE: bankruptcy and insolvency laws are standardly of public policy and cannot be contracted out. A key to damage control in the event of financial distress of a Joint-Venture Member is timeliness because preventive remedies are much more effective than curative remedies in such a context. Close and early attention to the bankruptcy or insolvency law of the country where the distressed Joint-Venture Member has its seat is very necessary.

14.13.2 Procedure of expulsion

Expulsion is done through a decision adopted by the Steering Committee. The concerned Joint-Venture Member must be given the opportunity to defend itself, but its presence in the meeting should not be required and it will not have a voting right. Expulsion shall become effective on the day of the decision of the Steering Committee, and it shall become binding if the Joint-Venture Member to be expelled has not, within 30 days of receipt of such decision, initiated Dispute Board proceedings in accordance with the JV Agreement or arbitration proceedings if the application of the DB has not been agreed upon. Expulsion cannot affect any of the existing rights or obligations of the Joint-Venture Member to be expelled, or any claims against it.

14.13.3 Continuation of the Joint Venture

If a Joint-Venture Member is expelled from the Joint Venture, then, of course, the Joint Venture continues with the remaining Joint-Venture Members.

14.13.4 Use of property

The Joint-Venture Members remaining in the Joint Venture should be allowed to continue to use the expelled Joint-Venture Member's physical property (such as equipment, materials, system components) and intellectual property (such as drawings and know-how), for which no substitute can be obtained at a reasonable cost and within due time. The expelled Joint-Venture Member must make every effort to facilitate such utilisation. The expelled Joint-Venture Member should be entitled to a reasonable compensation for such use by the other Joint-Venture Members. The compensation for such utilisation should only become payable once all accounts have been settled between the remaining Joint-Venture Members and the Employer.

The departing Joint-Venture Member shall have the duty without undue delay to:

- Make upon written request of the Project Manager a full disclosure of all information relevant for the continuation of the project, including disclosure of information regarding subcontractor(s) and subcontract(s) relevant to that Joint-Venture Member's contribution;
- Sign and execute all documents and perform all acts that the Joint Venture requires for the purpose of enabling the Joint Venture to recover any outstanding interest or right, or the property of or in connection with the Joint Venture or the remaining Joint-Venture Members; and
- Return to the Joint Venture or Joint-Venture Members all documents, records, papers or other property (in whatever form held) which are not owned by itself but may be in its possession or under its control, which relate in any way to the contribution or the Joint Venture's business affairs, and it shall not retain any copies thereof.

14.13.5 Liability for work provided

The expelled Joint-Venture Member(s) cannot be relieved from any of its liability for or in connection with its performance of its contribution up to the date of expulsion, subject to the limitations and/or exclusions of liability under the JV Agreement and/or the contract. The remuneration of the departing Joint-Venture Member for such supplies and services is to be paid in compliance with the tax regulations. The other Joint-Venture Members should be entitled to require such payments to the departing Joint-Venture Member on conditions as they think fit, including the provision of securities covering possible claims in accordance with the JV Agreement.

14.13.6 Liability for damages

To the extent the expulsion is due to causes within the control of the expelled Joint-Venture Member he is to be liable to the other Joint-Venture Members for damages (for example for the increase in cost to finalize the project and for liquidated damages payable to the Employer for delay to the extent they result from the expulsion).

14.13.7 Execution of expulsion by substitution

If expulsion of a JV Member is prohibited under the contract, and therefore the aforementioned procedure is not possible, then the remaining Joint-Venture Members could decide to reduce the contribution of, and where appropriate the liability of, the concerned Joint-Venture Member(s) to 0 (zero) and remove its voting rights in the Joint Venture. Where legal and practical the aforementioned liabilities of the expelled, or rather: substituted, JV Member are to apply *mutatis mutandis*.

14.13.8 Causes for termination of the Joint Venture

Upon a Joint-Venture Member's exercise of its right under mandatory law to terminate its participation in the Joint Venture, it shall leave the Joint Venture, and the remaining Joint-Venture Members shall continue the Joint Venture.

The Joint Venture should be automatically terminated in the following events:

1 Withdrawal of the invitation to bid by the Employer or announcement by the Employer that it will not award the contract to any of the bidding parties; provided, that if the Employer announces within a period provided for in the JV Agreement, say as an example 180 days, of such withdrawal or announcement of non-award that it will re-bid the project, then this Agreement shall remain in effect for the re-bid unless otherwise agreed by the Joint-Venture Members; or
2 The expiry of the period of validity of the Offer unless the Joint-Venture Members have agreed in writing to prolong the validity of this Agreement beyond such period; or
3 If the final Offer of the Joint-Venture Members is not accepted and no further negotiations are carried out between the Employer and the Joint Venture; or
4 When the Joint-Venture Members have unanimously so decided, and recorded such decision in writing; or
5 Failure of the Joint-Venture Members to agree on an Offer after negotiating in good faith for a reasonable time, and two Joint-Venture Members or the Joint Venture Project Management give the other Members e.g. five days' written notice of the intention to terminate the Agreement; or
6 In the event that the contract is awarded to the Joint Venture, once the contract has been performed or terminated and the Joint-Venture Members have fulfilled all their obligations and received their rights in accordance with the contract and Agreement.

14.13.9 Consequences of termination

The JV Agreement must provide for and the Steering Committee must decide on how to deal with:

1 Shared (or joined ownership of) assets acquired by the JV during the effectiveness of the JV Agreement;
2 The return of contributions made by the Joint-Venture Members by way of a right of use only, in both tangible and/or intangible assets which are to be returned to their Owner (and all licenses or leases, if any, shall be automatically terminated); and
3 The sale of any asset owned by the Joint Venture (any Joint-Venture Member may be a purchaser of all or any portion thereof), and the eventually remaining assets to be distributed to and among the Joint-Venture Members pro rata in proportion to their participation in this Joint Venture, without undue delay.

The termination of the JV Agreement may not affect any of the rights and obligations of the Joint-Venture Members accrued up to the date of the event causing the termination of the Agreement, including in respect of the sharing of profit and loss.

14.14 Claims, dispute resolution and arbitration

All disagreements and disputes must be resolved solely by the procedures set out in the JV Agreement.

If a disagreement arises out of or in connection with this Agreement during the time period of preparing, submitting and acceptance of the Offer, the respective Joint-Venture Members must attempt, in fair dealing and good faith, to settle such dispute amicably. The Joint-Venture Members must be free to organise the procedure of their settlement negotiations and to include senior management representatives as the case may be. Any of the Joint-Venture Members involved in the negotiation process may terminate the settlement negotiations at all times by written notification to the other Joint-Venture Member(s).

In order to avoid and resolve dispute(s) amongst the Joint-Venture Members, which can impede the continuation of the Works, the Joint-Venture Members should agree to establish a Dispute Board ("DB") in accordance with e.g. the ICC Dispute Board Rules in force at the date of signing the JV Agreement, which rules can be incorporated in the JV Agreement by reference. The selected DB should be established within 30 days after commencement of the project and should have one or three member(s) depending on the size and complexity of the project. Where the Joint-Venture Members have not otherwise agreed, there should be a Combined Dispute Board of three members.

All dispute(s) arising out of or in connection with the present Agreement which are not resolved in accordance by the DB must be finally settled under the Rules of Arbitration of the International Chamber of Commerce by one or more arbitrators appointed in accordance with the said Rules and the provisions of the JV Agreement. The place of arbitration and the language of the arbitral proceedings must be specified in the JV Agreement.

14.15 Choice of law

The JV Agreement must provide that it shall be governed by, and all disputes in connection with it shall be resolved in accordance with the substantive law of the country as specified in the Special Conditions to the JV Agreement without regard to this jurisdiction's conflicts of law provisions.

14.16 The Consortium

As the Consortium Members each have their own scope of work, rather than a joint scope of work as in a Joint Venture, and are towards the other members responsible for the correct execution of that scope of work and liable for failure of correct execution the members of a Consortium tend to keep things much more separated than in a Joint Venture.

14.16.1 The Steering Committee

In a Consortium the Supervisory Committee comprises delegates from each member who decide by unanimity. By Consortium Agreement the members may deviate from the unanimity in certain circumstances regarding the Works under the contract. E.g. the Owner wishes to instruct a variation that falls within the scope of only one member of the Consortium.

The Steering Committee may decide to appoint a Project Committee tasked to oversee the day-to-day execution of the project, similar to the Project Management in a Joint Venture.

14.16.2 The Consortium Leader

The Steering Committee may decide to appoint a Consortium Leader.

The duties of the Consortium Leader are of an often much needed administrative and coordinative character, including:

- To provide technical, commercial and organisational coordination of the Consortium Members in the bidding phase and during the performance of the contract;
- To act as the spokesperson in negotiations with the Employer, authorities or any third parties which are of joint interest, including preparing the necessary correspondence;
- To the extent applicable, to submit the bid for the Consortium and coordinate necessary formalities, in particular regarding a potential

mandatory registration of the Consortium, to keep the necessary books, and, to prepare and submit tax returns for the Consortium; to convene, preside and keep records of the meetings of the executive body(ies) of the Consortium;

- To make proposals for joint insurance coverage of all the Consortium Members;
- To coordinate the establishment of a joint construction site;
- To coordinate the preparation of progress reports and other documentation to be submitted to the Employer;
- If and as required, to keep an integrated information and digital documentation system and organise the sharing of such information and documentation with the Consortium Members and with the Employer as required;
- To collect data for invoicing the Employer and to collect payments from the Employer to the extent and in the way foreseen in the Consortium Agreement; and
- To perform any other duties assigned to the Consortium Leader by the Consortium Agreement or by a respective resolution or decision of the Steering Committee or of the Consortium Members supplementing or interpreting this Agreement.

14.16.3 Liabilities

Where in the Joint Venture the members share all rights, obligations, liabilities and risk in proportion to their participation in the Joint Venture, this is not so in a Consortium; the Consortium Members have recourse on a member for defects in the work within the scope of that member.

14.16.4 Expulsion

Unless prohibited by the contract with the Owner, a Consortium Member may be expelled from the Consortium. The grounds for expulsion and the consequences would be very much the same as described earlier for the Joint Venture.

14.17 Takeaways

Both Joint Ventures and Consortia need a strict management structure to manage the execution of the work that the Joint Venture or Consortium contracted to do with the Owner.

Ultimate power in the JV and Consortium is to be vested in a Steering Committee whose members make their decisions by unanimity, regarding strategic management of the JV or Consortium. The JV or Consortium Agreement needs a provision of what is to happen in the event of a deadlock in the voting; there will be a reconvening of the meeting a few days later after Steering Committee members have had the time to confer with their home organisations. The Steering Committee oversees the activities of the Project Management.

Project Management on sizeable projects is generally a three-person job with a Project Manager, an operations manager and a finance manager.

The participation of a member in a Joint Venture is allocated in proportion to their contribution to the JV; they share in profit and loss in proportion to their contribution to the JV. The establishment of each member's monetary value of contribution often causes a heated debate.

Consortium Members do not share in their joint performance of the contract with the Owner.

Consortium Members are liable towards each other for default in their individual scopes of work under the contract with the Owner. Both JV Members and Consortium Members share a duty of Good Faith and Fair Dealing towards each other. When working together, particularly on preparing the bid for a project they do so on an exclusive basis. No member can be involved in multiple biddings for the same project with different JV or Consortium partners.

INDEX